College Accreditation

College Accreditation: Managing Internal Revitalization and Public Respect

Jeffrey W. Alstete

College Accreditation: Managing Internal Revitalization and Public Respect
Copyright © Jeffrey W. Alstete, 2007.

Softcover reprint of the hardcover 1st edition 2007 978-1-4039-7420-4

All rights reserved. No part of this book may be used or reproduced in any manner whatsoever without written permission except in the case of brief quotations embodied in critical articles or reviews.

First published in 2007 by
PALGRAVE MACMILLAN™
175 Fifth Avenue, New York, N.Y. 10010 and
Houndmills, Basingstoke, Hampshire, England RG21 6XS.
Companies and representatives throughout the world.

PALGRAVE MACMILLAN is the global academic imprint of the Palgrave Macmillan division of St. Martin's Press, LLC and of Palgrave Macmillan Ltd. Macmillan is a registered trademark in the United States, United Kingdom and other countries. Palgrave is a registered trademark in the European Union and other countries.

ISBN 978-1-349-53479-1   ISBN 978-0-230-60193-2 (eBook)
DOI 10.1057/9780230601932

Library of Congress Cataloging-in-Publication Data

Alstete, Jeffrey W.
    College accreditation: managing internal revitalization and public respect/ Jeffrey W. Alstete.
        p. cm.
    Includes bibliographical references and index.

    1. Universities and college—Accreditation United States. I. Title.

LB2331.615.U6A45 2007
379.1'58--dc22   2006049472

A catalogue record for this book is available from the British Library.

Design by Macmillan India Ltd.

First edition: January 2007
Transferred to Digital Printing in 2011

This book is dedicated to my wife, Marta, and our daughter, Jessica

Contents

List of Tables and Figures	ix
Preface	xi
Acknowledgments	xv
1. Introduction	1
Part One: A Look at College Accreditation	9
2. A Brief History of College Accreditation	11
3. New Techniques for Accreditation Today	23
4. Accreditation under Fire	29
5. The Need for Accreditation Management	39
Part Two: The Accreditation Process	45
6. Getting Started on Accreditation	47
7. The Application for Accreditation	53
8. The Self-Evaluation Analysis	57
9. The Peer-Review Visit	67
10. Publication of the Evaluation	71
11. Following the Accreditation On-Site Evaluation	75
Part Three: Managing Accreditation at Colleges and Universities	79
12. Accreditation Management	81
13. Planning Accreditation	93
14. Organizational and Staffing Issues in Managing Accreditation	109
15. Controlling and Directing Accreditation	115
16. Writing the Self-Study Report	121
17. Budgeting for Accreditation Management	129

Part Four: New Strategies for Accreditation	131
18. Using Management Quality Techniques for Accreditation	133
19. New Electronic Tools to Support Accreditation Strategies	145
20. Accreditation of Distance Education	159
21. Conclusion	171
Appendix A—Recognized Accrediting Organizations	179
Appendix B—Nonrecognized College Accreditation Agencies	187
Appendix C—Accreditation Eligibility Requirements	199
Appendix D—Institutional Newsletter Content Prior to Accreditation Visit	203
Appendix E—Accreditation in the United States	205
Appendix F—Sample Guidelines for a Regional Accreditation Self-Study Report	221
References	231
Index	245

List of Tables and Figures

Tables

1. Historical Periods of U.S. Higher Education Accreditation   14
2. New Mexico Tech Self-Study Timeline   64
3. Connection of Regional Accreditation Criteria and College Strategic Planning   101
4. Linkage between Self-Study Challenge Areas and Strategic Planning   102
5. Stakeholder Groups for Baldrige Criteria   137
6. Baldrige Criteria Compared to NCA Criteria   138

Figures

1. Accreditation Turning Point in 2005   34
2. Need for Accreditation Management   42
3. Self-Study SWOT Analysis   59
4. Accreditation Management Framework for a Self-Study Planning Timeline   65
5. Accreditation Management and Institutional Goals   91
6. Unplanned Accreditation Time Schedule   95
7. Planned Accreditation Time Schedule Linked to Strategic Plans   96
8. Internet Support of Accreditation Management   154

Preface

New opportunities for students to continue learning beyond secondary school, increasing expectations, higher costs, and changing demographics are just some of the issues that are challenging higher education in the early part of the twenty-first century. The announcement that an institution is accredited is an important and worthy statement that a college or university is meeting the challenges that society presents today, and students are supposed to be encouraged that the years of effort and money spent on their education will be of value. Accreditation involves not only higher education, but also research institutions and elementary/secondary schools, which are also frequently undergoing accreditation reviews by a variety of associations, organizations, and government agencies today. It is a common complaint of faculty, administrators, staff, and trustees at these institutions that too much time, money, and organizational resources are spent on achieving and maintaining accreditation. One typical medium-sized institution recently reported that they have 17 accrediting agencies to comply with, and are continually under review by one or more such organizations every year. Faculty, administrators, students, staff, and other stakeholders are significantly involved in these important efforts when they serve on committees and task forces, or are appointed directors of accreditation. While several of the individual accrediting agencies do offer some printed guidelines, there is no comprehensive guidance available in the literature market today that offers strategic and operational strategies for managing accreditation processes in an effective manner. This book seeks to address this need by offering an informative resource on college accreditation in contemporary times, suggesting techniques for managing the accreditation process, and examining some models of best practices in accreditation management. It should be noted that there is one book on accreditation by the author (Alstete, 2004) that approaches accreditation using interpretive strategies with management of meaning and symbolic actions. The overall approach in this new book is focused on the more traditional linear aspects of accreditation management strategies. In addition to a different core approach, this book is updated and more comprehensive, including aspects of new technology implementation.

How can educational institutions manage internal revitalization and public respect of their activities? This book shows how management of academic accreditation through effective planning and implementation is part of the answer. Accreditation is examined and discussed by exploring the history, processes, current status, and potential future models of educational accreditation along with examples of successful management of accreditation at many postsecondary educational institutions. The management functions that are effectively being applied to accreditation oversight and implementation are organized into a taxonomy of lower elements such as controlling, operating, reporting, and budgeting, along with higher elements of planning, organizing, staffing, and developing the accreditation strategies for institutions. This classification of accreditation tools, concepts, approaches, and methods presents a unique viewpoint on accreditation today and offers practical guidance to both accreditation newcomers and experienced practitioners.

Audience

Readers familiar with the literature of higher education administration will recognize that this book follows the format of detailing the history, literature, and ideas behind the subject, followed by case studies of best practices, a model for implementation, and, finally, recommendations for institutions and educational leaders. The intended audience for this work includes college and university administrators at various levels, trustees, faculty involved in governance and accreditation matters, as well as faculty and students of higher education administration and leadership. Individuals in regional and specialized accreditation agencies, elementary and secondary school systems, and research institutions will also find this information helpful. In addition, anyone in the general public who is interested in learning about the important systems and methods in education oversight and improvement will find this book informative.

Overview of the Contents

This book is divided into four parts. Part one examines the background and history of accreditation in the United States, and how it is affecting the public perception about higher education and accreditation today. The needs and purpose of this book are also explored by briefly examining changes in accreditation that have taken place in recent years, along with the potential impact of issues affecting higher education, such as funding, changes in student/faculty demographics and priorities, distance learning, new instructional technology, and public perceptions. The evolution, common elements, and increasing criticisms by leaders in higher education and the public are also examined. Part two inspects the process of accreditation itself, and seeks to understand the steps involved in specialized and regional accreditation. The impact of accreditation on the institution and the related aspects of accountability and trust are explored with the purpose of creating a responsive and meaningful experience for the institution. The third part offers important guidance on the management of accreditation, along with examples of model programs at various colleges and universities. Issues such as the organizational structure at the institution, strategic planning, financial support, faculty leadership, use of information technology, team visit preparation, and overall guidance of the accreditation programs are discussed with the goal of providing thorough strategies for effectively managing the accreditation process. Other elements of accreditation management, such as planning, staffing, reporting, and budgeting are also studied. The recent implementation of distance-learning courses and programs and their inclusion in accreditation management processes are also examined. Finally, the fourth part offers a look at the new approaches to higher education, including quality-improvement strategies such as the Baldrige Award criteria, the impact of distance learning on accreditation processes, effective leadership strategies, and conclusions regarding the challenges that face the future of accreditation. These include potential reforms, academic audits, internal reviews, public accountability, and recommendations for process improvements both within institutions and accrediting organizations.

About the Author

Jeffrey W. Alstete is currently a full-time faculty member in the department of management and business administration in the Hagan School of Business at Iona College, New Rochelle, New York. Dr. Alstete earned a doctorate in Higher Education Administration from Seton Hall University (1994), and holds a Master of Business Administration (1987) and an MS in Computer Science (1990) from Iona College, as well as a BS in Business Administration from St. Thomas Aquinas College (1985).

Dr. Alstete has held several administrative positions at a number of institutions, including Associate Dean at Iona College, Director of Continuing Education at St. Thomas Aquinas College, and Assistant Dean for Graduate Programs at the W. Paul Stillman School of Business, Seton Hall University. In these positions, he served on and chaired regional and specialized accreditation task forces and committees, including a recent regional accreditation periodic review by the Middle States Association, and was appointed as Director of Accreditation at Iona College's Hagan School of Business during the final phase of their successful pursuit of initial accreditation by the Association to Advance Collegiate Schools of Business (AACSB) International.

He is the author of three books: *Benchmarking in Higher Education: Adapting Best Practices to Improve Quality* for the ASHE-ERIC Higher Education Report series (Jossey-Bass, 1995); *Post-tenure Faculty Development: Building a System of Improvement and Appreciation,* for ASHE-ERIC Reports (Jossey-Bass, 2000); and *Accreditation Matters: Achieving Academic Recognition and Renewal,* also for the ASHE-ERIC Reports (Jossey-Bass, 2004). This is a prestigious peer-reviewed issue analysis book series in higher education. He is also an award-winning author of journal articles related to higher education and business that have been published in scholarly journals such as the *Journal of Education for Business; Journal of Continuing Higher Education; Benchmarking: An International Journal; Team Performance Management;* and the *International Journal of Entrepreneurial Behavior and Research.* He is also the Book Review Editor for *Benchmarking: An International Journal,* and a reviewer for several other journals in higher education and business. Dr. Alstete's primary areas of scholarly research and consulting include administration effectiveness, accreditation, benchmarking, faculty development, distance learning, entrepreneurship, virtual teams, knowledge management, strategic management, and organizational improvement.

Acknowledgments

Just as an accreditation self-evaluation is a long process supported by many individuals with someone leading the effort, the work in this book had the support of several people and organizations. I want to acknowledge my colleague and former supervisor, Dr. Nicholas J. Beutell, for the many years of discussions we had together regarding the topic of accreditation. His experience as an accreditation consultant-evaluator for various colleges and universities provided a valuable perspective, and his insights about organizational challenges and model of integrity in educational endeavors were important to me in the creation of this book. I also want to acknowledge the thorough and very responsive services of the staff at the Ryan Library of Iona College. Several years of remote requesting of materials included many late-night e-mails, completing online request forms, telephone calls to the Reference Desk, and notifications of deliveries received from the library, which made the research portion of this book easier and more complete. In addition, I want to acknowledge several of the college accreditation associations and their directors for granting permission to use sections of their policies and documents as useful examples to readers.

Finally, the editor of this book, Amanda Johnson Moon and her editorial assistant, Emily Leithauser, were very professional in their support of the work on this book and deserve acknowledgment for their excellent service on behalf of their company, Palgrave Macmillan.

CHAPTER 1

Introduction

The field of higher education has witnessed dramatic growth in recent decades and is currently continuing on that path with added elements of change such as new technology, changing demographics, and increased accountability Parents and potential college students often compare and contrast the myriad of choices for higher education by reading magazine rankings, doing research on the Internet, asking friends and relatives, and sometimes considering regional and specialized accreditation. Employees in higher education ask other questions. We are performing effectively and many people already know this, so why do we need to pursue accreditation again? How can I properly conduct all of my regular duties, including teaching or managing my area in the institution, and also work on a tedious accreditation processes? These and other questions are often asked by administrators and faculty in educational organizations today. The answers should include the knowledge that accreditation is required for recognition by governmental agencies; that it is demanded by potential and current students, their families, and society; and that accreditation is developing into a more of a comprehensive system of self-renewal for institutions. Employees in this industry spend most of their professional lives involved with these complex systems of learning, and making improvements is not easy but should be a priority to ensure achievement, expansion, organizational advancement, and learning by the educational organizations.

There are many challenges facing higher education today, such as increased accountability and competition, shifting demographics, strained financial resources, maintaining relevant curricula, and changing technology. These factors are also helping to cause the accreditation system to undergo evolutionary development, and some of this transformation is related to the changes in the aforementioned challenges that educational systems face (Ewell 2001). A significant development is in the system or form of information transfer. It seems to some that computers and the Internet might be the basic reason for these developments; however, these factors are more facilitators of the developments rather than the actual cause. The means of knowledge distribution is evolving from the traditional faculty lectures and discussions, where the professor is assumed to have all the knowledge to impart to students, to new, effective, and active learning strategies where students take part in the method of learning with guidance and assistance provided by professors. Developments are also taking place in course and program models, with older 15-week term schedules becoming less common and new assurance of learning-founded systems becoming more widely used. Another development can be seen in student backgrounds, where, typically, students today attend several colleges and universities for their education, and come from diverse socioeconomic backgrounds. These developments in higher education and society are encouraging the evolution of institutional behaviors and faculty responsibilities to change in new ways. In regard to the institutional structures of higher education, changes in the centralization/decentralization of levels of authority (Alstete 1997) and in the accountability of budget authority to different units can affect the overall direction of the institution, as individual departments may be seeking divergent objectives that may or may not be in line with the mission of their institution. In the academic arena of faculty teaching and student learning, there are widespread changes in the use of active student learning and fewer traditional classroom-teaching practices by professors. There is increased emphasis on student team effort, student participation, faculty facilitation, and variety in teaching methodologies. Other trends include an increased emphasis on assessment of student-learning outcomes; changes in student ethnic, racial, and gender diversity; and the integration of new technology in the collegiate learning methods (Kezar 2000). People in higher education administration, faculty members, students, and employers of graduates are currently experiencing the consequences of these developments, and accrediting associations are now understanding and adapting accordingly.

Individuals, families, employers, and society expect a lot from higher education institutions in the United States. Such institutions are expected to provide valuable postsecondary education for traditional-age students and returning adults, and a variety of graduate programs. In addition, they are expected to provide a wealth of research for the benefit of humanity and even serve local populations with service programs— all of this for an economical price that a typical American family can afford. Naturally, the resulting conflict of these expectations with the reality of increasing costs, changing curricula, shifting demographics of students, strained financial resources, and evolving omnipresent technology has created a situation where the public's demands for accountability for quality and the ability of institutions of higher education to meet these challenges has created a conundrum. Part of the effect can be observed on the system, oversight, and proposed future of higher education accreditation. We are at a point beyond general and specific criticisms that are merely articulated by individuals, authors, and politicians. There is at least one proposal before the U.S. Congress that includes significant changes in the accreditation system and oversight of higher education. There are several potential outcomes that could result from these proposed changes, and we are therefore at a crossroads, or turning point, in accreditation and general public accountability in higher education today.

This can be seen as part of a larger issue involving the overall real or perceived changing role of higher education. Kezar finds that "according to critics, higher education is foregoing its role as a social institution and is functioning increasingly as an industry with fluctuating, predominantly economic goals and market-oriented values"

(2004 p. 430). Some of these aspects have benefited higher education and society, but they can also be viewed as compromising the long-term public and democratic interests that have previously characterized higher education. Among the results of these changes have been the effects on, perceptions of, and actions taken by the regional and specialized accrediting agencies of postsecondary education in the United States. We are currently at a crossroads in the relationship between society and higher education, particularly in regard to the area of self-regulation. For example, the U.S. government has recently been making attempts to override the peer-review system by proposing to specifically determine conditions of transfer credit, quality of distance-learning programs, and acceptable student-learning outcomes (Eaton 2004). This effort diminishes institutional autonomy in higher education, academic freedom, and the centrality of our mission. Regardless of whether the legal efforts currently under way in the federal government increase control and intervention, certain elements of society are undoubtedly concerned about the goals and quality of U.S. higher education. This is ongoing despite the generally positive public opinion about higher education.

In fact, a recent survey of 1,000 adults from 48 states by a marketing-research consulting company found that Americans' confidence in higher education is strong (Selingo 2004). Ninety-three percent of respondents agree that higher education is a valuable resource, and six in ten stated that four-year colleges in their states are offering high-quality programs. This perception has apparently not greatly influenced the elected state and federal officials and other critics, who are increasing skeptical. Therefore, there seems to be a disconnect between the general public perception and certain leaders, legislators, and outspoken critics of higher education with regard to what we in higher education are currently perceived as offering society. The truth is probably that there are indeed some problems that need to be addressed; however, American higher education continues to offer high-quality education, research, and related services to society.

The issues and questions arise again: Should we in higher education continue to police policy adherence and ensure quality? What are the purposes of accreditation? How frequently should the accreditation standards change? As far back as 1940, it was stated that "for many years suggestions have been made that the 'time may eventually arrive when the functions of accrediting educational institutions can be taken over entirely by some governmental agency, thus changing the process from an extra-legal to a legal control'" (Russell and Judd 1940, p. 159, cited in Selden 1960). Selden then states that there were/are dozens of refutations to this statement based on the political traditions in the United States. Historically, the purposes of accreditation include the desire to encourage institutions to improve, facilitate the transfer of students, inform employers of graduates about the quality of education received, raise the standards of education, and supply the general public with some guidance on which institutions to attend (Blauch 1959). These purposes are still strongly in place today, and perhaps even more so in light of the increasing tuition costs, strained institutional financial resources and public money availability, shifting demographics, and changing technology.

Change is ongoing not only in instructional technology, demographics, and the curriculum in higher education, but also in accreditation. However, exactly how frequently accreditation should change is an important issue for everyone concerned with higher education. Davenport (2001) states that accreditation should change "often enough to be responsive, but not so often as to be burdensome or intrusive" (p. 80). In addition, it is suggested that when changes do happen, broad-based audiences from within and outside higher education should be included in the revision process. Many of the existing policies and procedures used by accreditors today provide a controlled system for the continuous improvement of the change process to be assessed, and sensible revisions can then occur. Factors that were involved in the origins of accreditation, such as state governments, academic disciplines, voluntary and regional associations, and the federal government listing or statistical responsibilities (Harcleroad 1980) will also continue to impact the changing expectations for accreditation of higher education in the United States. However, from an international perspective, it has recently been perceived that American accreditors do not have explicit benchmarks or standards (El-Khawas 2001). The recent emphasis by specialized and regional accrediting agencies on outputs and outcomes has raised this question of explicit standards once more and makes change a continuing, ongoing factor.

Although extensive literature on specific accreditation processes exists, few attempts have been made to examine the current crossroads that accreditation is facing. There have been several books and monographs over the years that attempted to assemble a comprehensive body of information on the history and practices of accreditation and changes within (Alstete 2004; Blauch 1959; El-Khawas 2001; Harcleroad 1980; Lenn and Campos 1997; Selden 1960; Semrow et al. 1992), and these have often sought to treat systematically the various organizations and agencies concerned with accreditation in the United States. There have also been research journal articles published regarding various accreditation issues, including several that are devoted to exploring the issue of accreditation crossroads (Armstrong 1994; Barker and Smith 1998; Benjamin 1994; Davenport 2001; Dill et al. 1996; Doerr 1983; Glidden 1996, 2004; Pfnister 1971; Thompson 1993; Tobin 1994; Vaughn 2002; Vergari and Hess 2004). With this previous research in mind, this book attempts to start a discussion on the changes that accreditation is facing today and the issues that have led to the current crossroads, and proposes several directions for accreditation of higher education to take at this turning point.

In light of these changes, colleges and universities, as well as accreditation associations, are recognizing these developments and are now looking for new functions and models for their organizations in this increasingly complex milieu. Over 20 years ago, changes were anticipated and correctly forecast (Young et al. 1983). These include developments in accreditation approaches, from a highly numerical input-driven system to a more qualitative approach, where there is less reliance on quantitative data such as student performance rates and budget numbers, and more emphasis on the methods used for organizational improvement. There is also a movement by many accreditation associations away from a quest for college and university similarity toward distinctiveness. The accreditation agencies and their leaders now understand that the mission of the educational organization is to supply the direction for the accreditation review and not some association-driven standard. This shift is part of the general development in accreditation agencies away from the purely external examination of internal quantitative data toward more emphasis on institutional improvement systems, assurance of learning, and agency guidance of institutional development. Schools, colleges, and universities are now being required to perform more of the evaluation themselves and not rely so heavily on the external accrediting agency. Accreditation associations are now serving less as decision-oriented judges and more as catalysts, advisors, and counselors for guiding colleges and universities in their progress.

This evolution of the role of accreditation agencies is a needed development, because the process and standards for regional accreditation are perceived by many institutions as not demanding for most colleges and universities that have been accredited for many years (Dill 2000). Today, a significant question for institutions to ponder is, what will accreditation reviews include in the coming years? A potential analogy for educators to consider is the accreditation process as becoming more of an active-learning exercise for college and university organizations. Active learning in education has been generally defined as when students (learners) must do more than just listen to an instructor. They must be involved in working out problems and be dynamically engaged in such sophisticated thinking tasks as investigation, synthesis, and assessment (Bonwell and Eison 1991). In the management of accreditation and the actual process therein, this translates into a more lively involvement by colleges and universities in the accreditation self-study process, just as when active-learning teaching assignments include more engagement by student learners in their instruction, scholarship, and growth. College and university senior administrators can help faculty members, midlevel administrators, and students at their institutions to view accreditation review practice as a problem-based active learning exercise with tangible benefits. Certain institutional problems or challenges can be selected and examined by stakeholders in the institution and then reviewed by colleagues from other colleges and universities, and educated suggestions for addressing these challenges can be produced. During and after receiving this guidance, it is up to the senior administration of the institution under review to facilitate the perception of the accreditation review process as a helpful active-learning project and ensure that is positively received at the institution. However, strong, decisive leadership and management of the accreditation process, planning, and preparation must be exhibited by the senior faculty and administrators to ensure that these goals are achieved.

Administrative theorists (Gulick and Urwick 1937) have long held that management functions involve planning, organizing, supervising, directing, reporting, budgeting, and other management activities. In the field of higher education, these functions were not what many academics initially had in mind when they entered their profession and came into the field of higher education. Regardless of this ignorance or possible bias against administrative theory and practice, faculty and senior administration leaders can and should be solidly encouraged to effectively manage and contribute in systematic accreditation methods, for several reasons. First, self-regulation through the current system of specialized and regional accreditation offers colleges and universities the techniques and foundation to continue and reinforce educational reliability, organizational variety, and academic freedom (Benjamin 1994). A substitute for this self-rule is federal government-supervised licensing and approval of postsecondary educational institutions, which could interfere with the important collaborative features of the current educational oversight system that are still developing. The background, environment, attractiveness, and potency of colleges and universities in the United States and some other countries are founded in large part on the values of academic freedom, self-guidance, and independent intellectual spirit. It is for these reasons that it is critical for higher education leaders to inform faculty members and the community of educators about the merits of continuous evaluation by outsiders as their organizations are frequently scrutinized and examined by different accreditation associations.

The noted higher education scholar Martin Trow wrote that external reviews often choose simple numeric measures because they seem impartial and are easily accepted outside the institution (Trow 1994). However, in today's competitive and frequently changing higher education arena, institutions and their leaders should be constantly engaged in fundamental appraisals of their own actions and responsibilities. This is because the standards for accomplishment or excellence by outside assessments tend to be chosen with the more numerical and supposedly unbiased measures in mind, which unfortunately and regrettably can lead to a deterioration of institutional performance due to the larger quantity of technical information and the higher levels of efforts that colleges and universities are forced to put in as they adapt to the abridged trends of the numerical outputs. Trow explains that education units and the people who work there shape their behaviors toward what matters in the external accreditation reviews, which harms the soul of academe, because higher education is always more diverse and qualitative than any numerical summary assessment of inputs or results can evaluate.

There is a fairly recent movement by some thinkers in higher education who have maintained the argument that charitable self-regulation of colleges and universities is needed in postsecondary education to preclude direct and invasive federal and state agency oversight regulation, and also that the arrangement can be enhanced with an "academic audit" (Dill et al. 1996). An academic audit is defined as an "externally driven peer review of internal quality-assurance, assessment, and improvement systems" (p. 22). Unlike traditional education assessment, an academic audit is not performed to evaluate quality, but to concentrate on the procedures that are believed to generate quality and the systems by which educators are confident that quality has been accomplished. These audits of academic value usually occur at the overall institutional stage and focus on the official procedures of quality management such as policies, codes, rules, written guides, directives, and meeting notes. Despite this, academic audits do not look at areas such as academic standards, the condition of classroom instruction, or learning outcomes and assurance of learning (which many U.S. regional and specialized accreditation agencies now concentrate strongly on). However, academic audits do assess how colleges and universities satisfy their own standards and how these standards or goals are being accomplished. Additionally, academic audits have the advantage of shorter sequence times than traditional college accreditation review cycles. They are established on the belief that properly supported individuals working with adequate resources and high-quality methods will generate excellent outcomes and create unambiguous review tracks of accounts that were inspected. The method of self-governance and peer-oversight that most U.S. colleges and universities participate in must seek to balance and weigh the virtually contrary objectives of responsibility and development. The new proposed method of academic audits presents fresh alternatives and ideas for the traditional system of postsecondary accreditation to consider implementing. Whether systems such as accreditation or new academic audits are used, higher education leaders and participants would certainly prefer them to strict external federal government oversight and direction. Therefore, college and university leaders must scrutinize the features of accreditation (by reading this book and others) and seek methods and approaches to ensure that the accreditation system is a pleasing revitalization for the internal college community, is successful in improving the organization, and is acknowledged publicly as an essential practice and success that is richly deserved.

The general opinion about postsecondary education in the United States has been a concern in recent years, and a significant amount of investigation has been performed on public viewpoints toward colleges and universities (Eaton 1999). The inquiries have largely examined concerns related to the cost, value, reasons, and general worth of higher education, but have not focused on the public's familiarity with and attitude about academic accreditation. The Council for Higher Education Accreditation (CHEA) funded a survey in 1999 to discover what the general public understands or accepts as true about guaranteeing the excellence of education through the established system of accreditation. The survey findings revealed that a great preponderance of the public believe that higher education must achieve reasonable or elevated standards to earn accreditation. Interestingly, a high proportion (37%) of survey respondents did not know who conducts accreditation reviews. However, an apparent majority stated that they would not think of enrolling in a college course from a college or university that is not accredited. The results of this study are heartening for accrediting associations and colleges that are currently approved for accreditation. However, the study also reveals that there is a great need to continue to inform the public about exactly who the associations of accreditation are and the significant worth of peer-review systems that have worked successfully for many years. Moreover, the discrepancy between the general public's awareness of reasonable to elevated standards of accrediting bodies, and the internal opinion of the higher education industry (and the large proportion of colleges and universities that are accredited) concerning the allegedly low level of difficulty for attaining and maintaining accreditation approval, should be dealt with and scrutinized by all stakeholders concerned. This book on college accreditation seeks to offer background information and direction to both the general public and stakeholders in colleges and universities regarding the possible options for revitalization and public approval by successful accreditation methods and systems that are currently being used and some new ideas that are being proposed.

There are many latent paybacks to accreditation and self-study procedures carried out by higher education institutions and other types of educational ventures that are often not fully comprehended (Kells 1994). This may be due to the possible fact that many endeavors by educational members are troublesome, explanatory, perfunctory efforts that are not connected to the principal troubles and important accomplishments and openings that internal revitalization through accreditation can offer stakeholders and the public. Many constituents do state that some development results from accreditation processes and that they are useful endeavors, but the addition of nonstop self-study as part of ongoing continuous improvement policy in the organization is rarely attempted.

Instances of well-organized accreditation management processes are described later in this book, along with a discussion about how accreditation can be a useful and change-oriented process rather than a waste of valuable time, energy, and resources. College and university senior administrators and faculty leaders can stimulate employees of the postsecondary institutions to be energized about the accreditation process and instill faith that this process will revitalize the community and the organization as a whole. Some keys to success are: preparing the institutional community prior to the process, expecting more than minimal compliance with standards, planning beyond the traditional processes by exploring and including best practices (such as those discussed in this book), and keeping the continuous improvement outlook institutionalized and viewed as important by all the stakeholders involved.

PART ONE

A Look at College Accreditation

CHAPTER 2

A Brief History of College Accreditation

The English word accreditation is derived from the Latin word *credito,* which means trust. It is important for the public to trust that they are pursuing a worthy endeavor when a student embarks on a collegiate education, and the system to help ensure trust is what is called accreditation. In order to more completely understand college accreditation today, it is important to understand how it started, why it was begun, the historical development of academic accreditation that evolved over the years in response to society's changing demands, and its essential character therein (Young et al. 1983). After looking at the backdrop of accreditation and its chronicle, the valuable administration and development of accreditation systems can be more fully thrashed out in the framework of its ongoing development as a distinctive public and semigovernmental activity. The study of the chronological development of the evolution of colleges and universities in the United States is also important to fully comprehend the necessity for accreditation management within higher education today. Prior to the late nineteenth century, the curricula at colleges in the United States and the colonies was largely a proscribed doctrine based on an ancient course of study at educational institutions that were primarily religious in nature (Rudolph 1977). The curriculum normally included seven topics known as the "Trivium and Quadrivium." These were founded on the kinds of studies that were practiced in the classical world, and these seven liberal arts became codified in late antiquity by such writers as Varro and Martianus Capella (Fideler 1996). In medieval times, the seven liberal arts presented a canonical way of portraying the field of higher learning. The liberal arts were divided into the Trivium ("the three roads") and the Quadrivium ("the four roads").

The Trivium consisted of:

- Grammar
- Rhetoric
- Logic

The Quadrivium consisted of:

- Arithmetic—Number in itself
- Geometry—Number in space
- Music, Harmonics, or Tuning Theory—Number in time
- Astronomy or Cosmology—Number in space and time

The medieval Quadrivium thus followed the division of mathematics made by the Pythagoreans (Fideler 1996). Until the late nineteenth century, neither intercollegiate groups nor the government supervised the curricula and other matters. This started the movement toward a greater variety of higher education institutions, a greater diversity of student ages and previous learning levels, and overall educational quality.

A structure of official approval or regulatory supervision was probably not seen as needed because there were comparatively fewer colleges than today, only an insignificant number of the residents attended, and the set of courses did not cause unease to most people. However, as the nineteenth century came to an end, American universities were changing from a condition of very similar institutional types to one of individual variation and wide diversity (Veysey 1965). These important changes in growth and variability of colleges created situations where it was ineffective for the college professors to carry out the bulk of the institutional management, which they were used to doing, in areas such as student obedience and college admissions decisions in particular. While colleges such as Harvard established committees to deal with the bulk of the administrative duties, the variability and intricacy of postsecondary organizations shaped a requirement for consistency and conformity. Around 1890, a movement was begun to "accredit" the institutions that met minimal standards, which then became a major force after 1901.

Accreditation scholars have written that the first "period" in the historical development of accreditation began when the University of the State of New York (also known as the New York Board of Regents) was established in 1784 as a board for King's College in New York City (now called Columbia University) (Harcleroad 1980; Wilkins 1959). Interestingly, this system in New York was established after three years of squabbling, which is not unlike the current fervent discussions regarding accreditation issues and institutional control. The law was revised in 1787, allowing King's College and all other similar institutions to have their own boards of institutional authority. These conditions of institutional control and authority continue largely in the same form (Birch 1979; Orlans 1980; Selden 1960; and Wilkins 1959 in Harcleroad 1980); college "accreditation" activity therefore began at the state level in New York. Following this development and model, Iowa began oversight of colleges in 1846, Utah in 1896, Washington in 1909, Virginia in 1912, and Maryland in 1914 (Harcleroad 1980; Wilkins 1959). Concurrently, a system of voluntary, nonprofit, specialized associations seemed to have started with the American Medical Association (AMA) in 1847. Even though it created an early Committee on Medical Education, little effective control was applied until the beginning of the twentieth century, when the AMA was restructured. This development in the medical field unmistakably showed the feasibility of specialized accrediting agencies and how they could effectively monitor and approve quality of various specific programs in higher education in the United States.

In regard to periods of accreditation, several writers have offered historical perspectives on the stages or periods of accreditation evolution in the United States. Selden (1960) offers a detailed history of higher education curricula and accreditation in the United States from colonial times to 1960. He also writes that "opinions differ as to which organization first employed accrediting as a means of external control of educational standards" (p. 29) when he describes the history of several accrediting associations. The Board of Regents of New York State (NYS) is named as the most influential standardizing organization in existence and lays unquestionable claim to historical recognition being that its authority over higher education was enacted in 1787. Harcleroad (1980) concurs on the historical significance of the NYS Regents in the establishment of accreditation and marks that starting date as the beginning of the first historical period of accreditation. The periods of higher education accreditation have been labeled by several researchers (Alstete 2004; El-Khawas 2001; Harcleroad 1980; Semrow et al. 1992), with similar and somewhat overlapping categories for the various historical periods of development. These stages of accreditation development are shown in table 1.

The decades from approximately 1873 to 1909 were portrayed in the annals of education as an era that was illustrated by the growing dissatisfaction of many people in education with the typical college admission procedures and actions, and by the desire for rational interaction between secondary schools and the colleges of the that era (Shaw 1993). The primary topics of disorder were principally college admission standards, the suitable precollege preparatory subjects needed by college applicants, the proper function of the secondary schools, and turmoil regarding the increasing competition for college students (Tompkins and Gaumnitz 1954). Note that these issues are not unlike some problems still being addressed today in the early twenty-first century. Since the late nineteenth century, the federal government did not have the mandate to address the unanswered educational concerns that were outside the range of the state agencies, a development was then begun by participations of the higher education community in an attempt to resolve these concerns. It is important to note that rather than seek governmental supervision and guidance, the college leaders of the time sought to self-regulate. This may be due in part to the traditionally restricted function of the federal government in American society (as designated by the founding fathers) and the traditionally strong beliefs of individuals in self-reliance, unlike those in many other countries.

Primarily, the accreditation evolution sped up when it became known as a national happening on August 3–4, 1906, at a gathering of the joint committee of the National Association of State Universities in Williamstown, Massachusetts (Young et al. 1983). The goal of the get-together was to create a plan for building, protecting, and approving a common conception of admission standards and related management of the principles. The gathering was attended by representatives of the four existing regional associations, along with the newly created College Entrance Examination Board (CEEB). At the meeting, it was agreed to recommend that the regional associations have their member colleges accept certificates from accredited secondary schools in other regions. The attendees also recommended advancing the

Table 1 Historical Periods of U.S. Higher Education Accreditation

Alstete (2004)	El-Khawas (2001)	Harcleroad (1980)	Semrow (1992)
1880s–early 1900s: Focus on admission standards, definition of postsecondary institutions	1920s: College's structure and programs, # library volumes, classrooms, budget, very quantitative.	1787–1914: Starts with NYS Regents up until the removal of the Association of American Universities List	Late nineteenth century: The beginnings
1900s to early 1970s: Attempts at national coordination of agencies, more specialized agencies, input-driven standards	1930s–1940s: Single set of standards dropped, more holistic, greater flexibility allowed for interpreting standards	1914–1935: More qualitative and less quantitative, expansion of specialized agencies	1890s–1930s: Accreditation takes form, period of inspection blanks
	1950s: Rapid growth of higher education, accreditation served as screening mechanism before a review is allowed	1935–1948: Listing form was terminated, accreditation based on institutional mission is allowed	1930s–1950s: Transforming the basis of accrediting, *Manual of Accrediting*
1970s to the present: Diversity of quality standards among regional and specialized agencies, other models for periodic review allowed, increasing criticisms of the system	1980s: Change in accrediting agencies from emphasis on inputs to actual results.	1948–1975: Major changes, increased enrollment by returning veterans leads to accreditation linkage with federal funding	1950s–1970s: Continuing transformation, end of the *Manual*, beginning of the *Guide*
	1990s: Requirements became more demanding, focus on outcomes assessment	1975–1980: New COPA attempts to stem proliferation and stop federal government efforts to increase oversight	1970s–1980s: *Attempting a New Synthesis, The Interim Guide*, and *Handbook of Accreditation*
Early twenty-first century: Future may include increased accountability and quality of reviews	Late 1990s: Move to streamlining, simplification, educational effectiveness and outcomes		1990s and Beyond: Next steps, accreditation process and purpose still in question

arrangement of a college entrance certificate board or a commission for accrediting schools, to cultivate common definitions and standards, and to found a permanent commission on entrance requirements. These were practical matters related to college admissions and intercollegiate collaboration for standardization and mutual acknowledgment. As the total number and variety of postsecondary colleges was growing, these early accreditation issues were a natural part of the evolution of the multifaceted system of higher education that we see today in America.

The two primary types of accreditation that we see today—regional accreditation (for individual institutions) and specialized accreditation (for specific programs)—had their beginnings during this early period of accreditation development (Young et al. 1983). One of the first leaders in the regional accreditation arena was the North Central Association (NCA), which accredited secondary schools in 1905. The association then began accreditation or formal approval of its member colleges, and developed a set of college accreditation standards in 1909. The first published list of accredited institutions appeared in 1913 (Pfnister 1959; Young et al. 1983).

In the area of specialized (or professional) program accreditation, many examples of important early accreditation initiatives were seen in the health care field, where medical schools were an early adaptor of this concept. The significance of determining educational standards and quality assurance rapidly became obvious to the leaders of medical education and the public in America around the beginning of the twentieth century (Hamm 1997). As mentioned earlier, the AMA Council on Medical Education was formed in 1904 to look into those quality issues in medical education in the United States. A commissioner from the Carnegie Corporation prepared a report on the necessity for common standards in U.S. medical schools, at the request of the AMA Council. A decisive report was published shortly thereafter in 1910 and guided the creation of a national accrediting system for medical schools. As a result of this, many medical schools in the United States were subsequently closed after the establishment of those standards, and the accreditation agency (now called the Liaison Committee on Medical Education of the AMA) has had a large influence on the whole health care system in America.

Besides professional programmatic medical education, the Carnegie Foundation was influential in the standardization of postsecondary or college-level education in the early 1900s. For instance, in 1905, the wealthy industrialist Andrew Carnegie established the Carnegie Foundation for the Advancement of Teaching and funded this organization with a starting amount of ten million dollars (Shaw 1993; Tompkins and Gaumnitz 1954). This capital, along with the generated income, was directed to be invested to financially support the retirement of faculty members in America and Canada. As the Foundation began its task, the board of trustees of that organization soon learned of the wide variety of and lack of conformity among college education faculty and institutions in North America. A declaration was then made by the Carnegie Foundation that in order to appropriately achieve its objectives it must determine what rightly characterizes a college education and the principles of value and definition for colleges of that time. The Foundation, however, was a private association that did not have the power to compel colleges to meet the terms of the policies and regulations established by its board of trustees. Nevertheless, if any college and faculty members employed there desired to obtain any of the retirement funding then they were to conform to the stated Carnegie standards. The outcome of this was a set of minimal standards that contained a classification of what institutions of postsecondary education must have in regard to faculty, courses, and admission requirements. The phrases "Carnegie unit" and "Carnegie classification" are still commonly used today. Even though the effect of the Carnegie Foundation on the standardization of higher education was not thought of or precisely described as accreditation, it undoubtedly influenced colleges in

North America greatly in their development and paved the way for ensuing accreditation enactment by academic and professional associations.

In much of the world outside the United States, the establishment and maintenance of most educational standards are usually the responsibility of government agencies (Shaw 1993). However, many of the regional and specialized accrediting agencies that were founded in the United States now accredit programs internationally. In the United States, primary and secondary education is of "special interest" to the federal government, but it is the legal responsibility of the state governments. The 50 state governments normally grant control of the education programs to the thousands of local boards of education. Postsecondary education is also authorized by many of the state governments, with recognition currently granted by the Council for Higher Education Accreditation (CHEA) primarily through the regional accrediting agencies. After several years of examination with regard to the role of the U.S. government's predecessor agencies (the Council on Postsecondary Accreditation [COPA] and the Commission on Recognition of Postsecondary Accreditation [CORPA]), the CHEA was formed in 1996 following an extensive and searching debate about the appropriate role for a national organization concerned with accreditation of higher education institutions. The official position on this arrangement is that the presidents of U.S. universities and colleges established CHEA to strengthen higher education through strengthened accreditation of higher education institutions (CHEA 2001). As its mission statement provides, "The Council for Higher Education Accreditation will serve students and their families, colleges and universities, sponsoring bodies, governments, and employers by promoting academic quality through formal recognition of higher education accreditation bodies and will coordinate and work to advance self-regulation through accreditation." The office states that CHEA carries forward a long tradition that recognition of accrediting organizations should be a key strategy to ensure quality, accountability, and improvement in higher education. Recognition of accrediting organizations by CHEA affirms that standards and processes are consistent with the quality, improvement, and accountability expectations that CHEA has established. CHEA currently recognizes regional, specialized, national, and professional accrediting organizations. Pending development of a CHEA recognition policy and procedures and CHEA review of an accrediting organization's application for recognition, CHEA honors recognition provided by predecessor U.S. government agencies: the COPA and the CORPA. (For a complete list of accrediting organizations that meet CHEA eligibility standards, were recognized by COPA or eligible for recognition by CORPA, or are recognized by the U.S. Department of Education (USDE) or both, see Appendix A.)

CHEA states that recognition shall be understood to convey only that the organization meets CHEA's standards. Such recognition is not in any way intended to infringe on the right of any academic institution to determine for itself whether it should or should not affiliate with any accrediting organization. The U.S. federal government, through the USDE, also recognizes accrediting organizations. Federal, as distinct from CHEA, recognition aims to assure that the standards of accrediting organizations meet expectations for institutional and program participation in federal initiatives, such as student aid. State licensure reviews, too, serve important public purposes, including consumer protection in the higher education field.

Regional accreditation in the United States through the predecessors of CHEA came into being as public recognition and monitoring needs arose (Bloland 2001). The effort to arrange regional accrediting actually originated from secondary school officials in the New England and North Central regions, and for postsecondary collegiate level institutions first in the Middle and Southern regions of the United States. The Middle States Association of Colleges and Schools (MSACS) was founded in 1887; the aforementioned North Central Association of Colleges and Schools (NCA) and the Southern Association of Colleges and Schools were two regional associations that were founded in 1895. As American development moved westward, the Northwest Association of Schools and Colleges was formed in 1917, followed by the Western Association of Schools and Colleges in 1962. As the years progressed, these regional accrediting agencies differed greatly in size and influence, as they still do. Today, there are six regional accrediting organizations in the United States. Each accredits a total of well over 3,000 educational institutions, with individual regional agencies having between 149 (Northwest) and 960 (North Central) institutions (CHEA 1999). In looking at the various websites of the six regional accrediting agencies, some similarities in goals and objectives can be seen, but there are also large differences in methods, strategies, support, and organizational style. Despite these differences, this book will explore effective practices for helping to successfully manage regional and specialized accreditation from the perspective of a typical college or university that is striving to achieve public recognition and internal renewal of their institution.

Many educational institutions seek accreditation by regional and specialized accrediting agencies for public recognition, government approval, and to improve quality. In addition, the American specialized and regional accrediting agencies are now accrediting institutions outside the United States. As of 1999, the eight regional commissions (some of the six regional agencies have two commissions) reported that they were accrediting 160 institutions or programs, almost all of which were operating outside the United States (Bloland 2001). Calls for regional accreditors to expand their international review activities are increasing, and the federal government is giving more attention to international higher education. This is because institutions around the world now understand the value of the U.S. regional accreditation, the quality and value of these reviews, and the effect of distance-learning operations on the marketplace. As we shall explore, accreditation of distance learning is a complex and important matter of increasing importance to the government, educational institutions, and the public at large. Regional accreditation is the backbone of quality assurance in the United States and, increasingly, around the world. It establishes and sustains public confidence in the education system, and offers guidance for making improvements in the higher education.

The distinctiveness of accreditation and its rationale have been developing over the years, and they continue to change as society brings new challenges to higher education. However, some features have not altered much, such as the important and somewhat unique voluntary nature of the process, and the self-regulatory factor, where institutions are trusted with creating and enforcing standards. It is of great consequence to note that institutions of higher education are not required to seek accreditation and that the process is voluntary, to an extent. It has been written that accreditation is a voluntary private-sector requirement to the extent that any institution can choose to not seek it (Young et al. 1983). However, institutions that do not achieve accreditation are greatly disadvantaged in areas such as peer recognition, public perception, and funding support. The most crucial disadvantage in not achieving regional accreditation in the United States is that students enrolled in nonaccredited colleges or universities are not entitled to federal or many state-sponsored student-aid funds, and the general public perception of nonaccredited colleges is usually (and rightly) poor. It is remarkable that this multifaceted, systematic, and essential structure of education accreditation is a case model of a service industry that is self-regulated and not directly administered by a government agency. Therefore, leaders and employees of colleges and universities would be smart to remember this as they perform the duties of accreditation self-studies, because an option such as direct government agency control could be a lot more challenging and not nearly as successful in managing quality improvements. The latest incarnation of accreditation methods in the United States focuses largely on evaluating educational value through institutional self-evaluation, and provides direction or advice for improvement from an external viewpoint that is linked to the institution's strategic goals and overall mission. Certainly there are many who criticize the system (as the reader shall learn shortly), but it also has many beneficial features, is surprisingly robust, and somewhat unique in institutional quality management and improvement of educational processes.

In the late nineteenth century, the most important concerns in higher education oversight were the variability between a secondary school and a college, and of course the homogeneity of college admission standards with regard to secondary school expectations. At this time when the regional accreditation associations were establishing themselves, there was also a development in the creation of principles that could be practical across America in colleges (Bloland 2001). There were a variety of problems that needed to be addressed, such as the transfer of educational credit between colleges, the admission criteria for graduate schools entrance, and the decision regarding the equivalency of degrees between the United States and Germany. When combined, these concerns created the impetus for the national standardization of collegiate education in America. In today's large world of higher education with established systems in place, is difficult to see how these somewhat routine concerns were of national importance to colleges and universities. Accreditation purposes then were largely driven by inputs, resources, procedures, and administrative protocols. In the first half of the twentieth century, some of the organizations that were active in national accreditation-type movements were the Association of American Universities, the aforementioned Carnegie Commission for the Advancement of Teaching, and the American Association of University Women. By the middle of the twentieth century (1949), a national organization on collegiate accreditation was established: the National Commission on Accreditation (NCA). However, the regional accreditation associations were already well recognized and accredited most of the colleges that were already established. Not surprisingly, the regional accreditation associations were fundamentally not in favor of joining or being forced to join a national accreditation system (Bloland 2001). Subsequently, regional accreditation associations then formed the National Committee of Regional Accrediting Agencies (NCRAA), with the goal of assisting their members and to address the issue of the explosion in the number of national accrediting agencies.

The development then continued as the NCRAA became the Federation of Regional Accrediting Commissions of Higher Education (FRACHE), representing the eight postsecondary commissions of the six regional associations. The FRACHE was seen as not performing up to its capability and did nothing to stop the growth of professional and vocational accreditation agencies across the United States.

It was at this time that the higher education accreditation leaders solidified their desire to respond to the comparable yet somewhat opposing interests of the federal government, state governments, and the increasing number of accrediting agencies. In fact, the presidents of many accredited colleges and universities became so dissatisfied with the difficult accreditation system that by 1975 a new national organization was created by integrating the NCA and the FRACHE into COPA. This organization included a large variety of institutions in higher education, including community colleges, liberal arts colleges, proprietary schools, graduate research programs, religious schools, and trade and technical schools (Chambers 1983). Since its inception, COPA had configuration and managerial problems primarily because there was no official position for the staff employees. The accreditation questions it was involved with clashed with their supply of money from the accreditation members for whom the association was also making decisions regarding institutional accreditation. The rationale, extent, authority, and organization of COPA were frequently challenged, and many inquiries were made as to whether the association was even needed (Bloland 2001). A proposal for disbanding the association was produced by a special committee in April 1993, and this resulted in the creation of the CORPA to carry on the duties of officially approving education accrediting associations for the U.S. government and other agencies. Following this and after several more years of discussion, government control changes, and feedback from various groups, CHEA was officially formed in 1996 and is designated as the successor organization to COPA and CORPA for U.S. accreditation oversight.

Today, CHEA recognition of accrediting organizations has three basic purposes:

TO ADVANCE ACADEMIC QUALITY. To confirm that accrediting organizations have standards that advance academic quality in higher education; that those standards emphasize student achievement and high expectations of teaching and learning, research, and service; and that those standards are developed within the framework of institutional mission.

TO DEMONSTRATE ACCOUNTABILITY. To confirm that accrediting organizations have standards that ensure accountability through consistent, clear, and coherent communication to the public and the higher education community about the results of educational efforts. Accountability also includes a commitment by the accrediting organization to involve the public in accreditation decision-making.

TO ENCOURAGE PURPOSEFUL CHANGE AND NEEDED IMPROVEMENT. To confirm that accrediting organizations have standards that encourage institutions to plan, where needed, for purposeful change and improvement; to develop and sustain activities that anticipate and address needed change; to stress student achievement; and to ensure long-range institutional viability. (CHEA 2001, p. 572)

Where early accreditation strategies were organized merely to recognize what defines a college and standardize the admission policies, these purposes are clearly well evolved beyond those first goals, and the latest agenda is to ensure that the regional and specialized accrediting associations have more comprehensive goals. The CHEA also wants accrediting agencies to employ suitable and balanced measures in accreditation decision making and frequently reassess their accrediting practices. It has been summarized to state that the CHEA wants accrediting associations "to practice what they preach" with their education institution clients. However, there are opponents to the current organization of accreditation oversight by the CHEA, who see this system as a concession to college and university leaders instead of demanding thorough and high-quality methods in the supervision of the set of common standards to be used by the regional accreditation associations (Amaral 1998). At this still fairly early point in the twenty-first century, it remains to be seen if the foundation of the CHEA will be enough to protect the institutional self-government valued by the colleges and universities from any additional federal and state interference, especially in light of the very recent challenges to the accreditation system that will be explained shortly.

While any college or university can assert that it is "accredited," the use of this term is not truly regulated by associations or governmental agencies (Degree.net 2000). Anyone concerned with a precise meaning for the word can look in a dictionary and see that the term has as one of its meanings "to certify as meeting a prescribed standard" (Davies 1981). The standard(s) to be met by colleges and universities seeking to achieve accreditation are certain criteria and requirements that are published by the various accreditation agencies. However, some readers may wish to know whether an institution or accreditation agency is legitimate:, and there are several ways to verify this. One is to look at the various lists: in the appendix of this book, the CHEA, the USDE, or appropriate respected agencies in other countries. There has even been an attempt by some to create simple guidelines to define whether or not an educational institution can be considered to be accredited by an official agency recognized under Generally Accepted Accrediting Principles (GAAP) (Degree.net 2000). This acronym is borrowed from the field of accounting, where the Generally Accepted Accounting Principles are specific standards to which professional accountants can be held. The proponents of the accreditation GAAP concept in the United States believe that there is near-unanimous agreement on the concept (yet it may be called by different names). The relevant key decision makers are higher education registrars and admissions officers, corporate human resources officers, and government agencies. It should be noted that in some countries, the term "accreditation is not used," although that country's education evaluation procedure (such as the British Royal Charter) is accepted as "accredited" under GAAP.

For an agency or association to offer legitimate collegiate accreditation under GAAP, an accrediting agency must meet at least one of the following four criteria: (1) It must be recognized by the CHEA in Washington, D.C.; (2) It must be recognized by the U.S. Department of Education; (3) It must be recognized by (or more commonly, a part of) their relevant national education agency; or (4) Schools they accredit must be routinely listed in one or more of the following publications: the *International Handbook of Universities* (a UNESCO publication), the *Commonwealth Universities Yearbook,* the *World Education* Series, published by American Association of Collegiate Registrars and Admission Officers (AACRAO) - International Education Series, and/or the *Countries* Series, published by the National Office of Overseas Skills Recognition in Australia (NOOSR). (Degree.net 2000 p. 881; NOOSR 2004, p. 885; Palgrave 2005 p. 886; Palgrave 2005 p. 887).

To many faculty and administrators, accrediting agencies today are often viewed as consultants to colleges for serious quality improvement of educational organizations, with the potential to achieve great changes if properly utilized. The plethora of regional, specialized, and international accrediting agencies is still a challenge for colleges and universities, so a sensible, practical, yet far-reaching model of organizational process planning is needed by faculty, administrators, and educational leaders to meet the many demands of various accrediting agencies to continually improve. As we shall discover, many colleges and universities are leveraging their accrediting efforts in a manner that encourages positive changes to long-held practices, recognition of achievement within and outside the organization, and even improvement in the traditional amount of self-evaluation undertaken, where earlier there was limited or no effective internal management of accreditation efforts at their institutions.

CHAPTER 3

New Techniques for Accreditation Today

Quality improvement initiatives that were developed in the field of business have been adapted by many educational institutions (Alstete 1996a). Techniques such as the balanced scorecard, benchmarking, key performance indicators, total quality management, business process reengineering, management by objectives, zero-based budgeting, and other techniques have been and are currently being used by postsecondary institutions today to help address the demands for continuous improvement. While the original purposes of higher education accreditation evolved during the twentieth century to meet the changing needs of the public, there were also tremendous increases in the number, size, and diversity of colleges and universities. Demands for increased accountability in higher education escalated in the late 1960s and early 1970s as the costs for government regulation and reporting of many college functions increased. These increased accreditation costs for institutions of higher education, along with general economic inflationary strains on institutions, forced an offloading of these expenses to students. The results yielded an increased scrutiny by the public and others of the answerability and the function of accreditation of colleges and universities. One example was a report in 1979 by COPA that called for "accreditation teams to begin to look for evidence of student achievement (outcomes) used for the award of credit and degrees, and make judgments about the quality of the institution in light of the adjudged student achievement compared with degrees awarded" (Casey and Harris 1979, p. 25). This was one of the initial calls for outcomes assessment, which became ubiquitous in many accreditation standards and sustained the development of what the function of accreditation was to become in later years. Overall, the self-governance of accreditation methods and strategies were changing to become more than just self-protective statements against gradually more pointless accreditation standards.

Aside from costs to students and their parents, there was a strong demand for accountability and higher quality of graduates from the employers of college and university alumni in the business world and other sectors, such as health care and education. These factors, combined with higher college tuition costs and congested college classrooms at many large public institutions led to a heightened awareness of education value issues in postsecondary preparation that needed to be addressed. For the business world, global competition helped create a large movement to increase quality of American goods and services. An example of this is the Malcolm Baldrige National Quality Award that was created by public law in 1987 (NIST 1987). The Baldrige Award program led to the creation of a new public-private partnership, and also included a special category for educational organization award winners. Principal support for the program comes from the Foundation for the Malcolm Baldrige National Quality Award, established in 1988. The award is named after Malcolm Baldrige, who served as secretary of commerce from 1981 until his tragic death in a rodeo accident in 1987. His managerial excellence contributed to long-term improvement in efficiency and effectiveness of government. The creators of this award wrote that:

1. The leadership of the United States in product and process quality has been challenged strongly (and sometimes successfully) by foreign competition, and our Nation's productivity growth has improved less than our competitors' over the last two decades.
2. American business and industry are beginning to understand that poor quality costs companies as much as 20 percent of sales revenues nationally and that improved quality of goods and services goes hand in hand with improved productivity, lower costs, and increased profitability.
3. Strategic planning for quality and quality improvement programs, through a commitment to excellence in manufacturing and services, are becoming more and more essential to the well-being of our Nation's economy and our ability to compete effectively in the global marketplace.
4. Improved management understanding of the factory floor, worker involvement in quality, and greater emphasis on statistical process control can lead to dramatic improvements in the cost and quality of manufactured products.
5. The concept of quality improvement is directly applicable to small companies as well as large, to service industries as well as manufacturing, and to the public sector as well as private enterprise.
6. In order to be successful, quality improvement programs must be management-led and customer-oriented, and this may require fundamental changes in the way companies and agencies do business.
7. Several major industrial nations have successfully coupled rigorous private-sector quality audits with national awards giving special recognition to those enterprises the audits identify as the very best; and
8. A national quality award program of this kind in the United States would help improve quality and productivity by:
    a. helping to stimulate American companies to improve quality and productivity for the pride of recognition while obtaining a competitive edge through increased profits;
    b. recognizing the achievements of those companies that improve the quality of their goods and services and providing an example to others;

c. establishing guidelines and criteria that can be used by business, industrial, governmental, and other organizations in evaluating their own quality improvement efforts; and
   d. providing specific guidance for other American organizations that wish to learn how to manage for high quality by making available detailed information on how winning organizations were able to change their cultures and achieve eminence. (NIST 1987)

It is clear that the overall tone and intent of this kind of thinking with regard to quality improvement has affected the push for increased accountability and quality in higher education by political leaders, business managers, and the public. Among the five winners of the 2001 Malcolm Baldrige Awards, three were educational organizations (NIST 2001a). This was the first year that winners were named in the education category. Any for-profit or not-for-profit public or private organization that provides educational services in the United States or its territories is eligible to apply for the award in the education category. This includes elementary and secondary schools and school districts, colleges, universities, and university systems, schools or colleges within a university, professional schools, community colleges, technical schools, and charter schools. Applicants must show achievements and improvements in seven areas: leadership; strategic planning; student, stakeholder, and market focus; information and analysis; faculty and staff focus; process management; and organizational performance results.

All applicants for the Baldrige Award undergo a rigorous examination process that ranges from 300 to 1,000 hours of outside review. Final-stage applicants are visited by teams of examiners to clarify questions and verify information. All applications are reviewed by an independent board of examiners primarily from the private sector. Each applicant receives a report citing strengths and opportunities for improvement. The Baldrige program is managed by the National Institute of Standards and Technology (NIST), an agency of the Department of Commerce's Technology Administration, in conjunction with the private sector. It is similar to accreditation in that it can be considered a "quasi-governmental" program that is managed by the private sector but with some oversight by the federal government. Later in this book, we shall examine how processes such as the Baldrige program can be part of an overall self-evaluative quality improvement strategy that manages accreditation with other programs such as this one.

There have also been additional quality improvement strategies adapted from the business world, aside from the Baldrige Award, these include programs such as the Academic Quality Improvement Project (AQIP) that is run by the Higher Learning Commission of the NCA, which is a regional accrediting agency in America (AQIP 2002). The AQIP uses a quality system for self-evaluation that validates an educational organization's dedication to a disciplined or methodical quality-enhancement approach. This self-study permits the college or university to scrutinize its strong point as well as the areas that need improvement (weaknesses) using a predecided list of measures or questions that are based upon quality principles. Participating institutions may utilize the AQIP Criteria, other standards such as Baldrige, a response report from one of the many state quality programs, a report from an independent consultant appointed by the education institution, or from a procedure such as the Continuous Quality Improvement Network's (CQIN) Pacesetter Award program (CQIN 2003). The AQIP self-assessment is normally broad-based and evaluates the entire organization using the outside measures from these or other sources. Illustrations of colleges and universities that use this approach are shown in a later chapter.

There are also other examples of how accountability by corporate managers is becoming more important in light of the ethical issues surrounding corporate scandals of the early twenty-first century. Back in 1993, Peter Drucker wrote in his book *Post-Capitalist Society* that there will be demands for a "business audit" as a widespread means for performance verification at corporations to evaluate if their accomplishments are in line with the companies' strategic objectives (Drucker 1993). As mentioned earlier, educational institution audits, in comparison with conventional accreditation processes, have also been proposed for higher education reviews (Amaral 1998). In addition, some of the most recent regional and specialized accreditation standards call for audit-type evaluations in which demonstrated performance by institutions must be linked directly to the organization's strategic and operational plans. Though not clearly articulated in the higher education literature today, the business quality movement has been influential, in small and large ways, on higher education institution's internal planning and external reporting.

Where the traditional American accreditation process ascertains whether a college or university organization or specific program has met specific quality standards and minimal criteria, the European Association of Universities (CRE) has developed a structure of quality inspections (Amaral 1998; Dill 2000; Dill et al. 1996). While established academic accreditation evaluates the pragmatic functioning of education institutions with agreed-upon measures that are usually determined by the accrediting agency, an academic audit is an outwardly driven peer appraisal of internal institutional quality assurance, measurement, and quality-improvement systems. However, an audit does not evaluate the quality outcomes or results like an assessment, because an audit focuses on the institutional processes that are supposed to fabricate the quality and the actual systems that the education institutions can use to guarantee that quality is being attained. The size of the review team is not large in the CRE system, normally consisting of just three auditors. The team usually visits the college or university twice, first for a preliminary visit and then a primary visit, with the objective of attaining a formative decision about the quality management and strategic organization capabilities of the university. Nonetheless, the European system of audits has been criticized for being deficient in follow-up support after the audit review has been completed and the need for recurring follow-ups of assessed institutions, as in traditional collegiate accreditation systems (Amaral 1998). Additionally, due to the varied nature of the societal cultures among the member countries in the CRE, there is no solution to the recurring difficulty of improving clarity and similarity among the members of the CRE. American colleges and universities have expressed similar concerns regarding the different assessment and oversight systems that have been put in place by the varied regional and specialized accrediting associations. There is apprehension that these review methods may trail rather than guide the development in quality-improvement strategies due to the friendly and collegial spirit of these types of evaluations (Moore and Diamond 2000;

Trow 1994). Some see this as an explanation of why, after World War Two, most postsecondary education organizations started to implement a research-institution model instead of the teaching model that many of the colleges and universities were founded on. But we shall see that the accreditation system is today evolving into an improved quality-assurance system that now encourages adherence to the educational institution's roots and mission, and even supports multiplicity of process types and results, provided the outcomes are high in quality.

There have also been attempts in the corporate sector to attempt accreditation-type systems for quality improvement. One example can be seen in the travel agency industry, when a corporate travel department accrediting program was launched in 1998 by the Airlines Reporting Corporation (ARC), but had a slow start in gaining acceptance (Michels 1998). The ARC is a service company owned by the principal scheduled airlines of the United States. One of its primary functions is to process and evaluate applications by persons and organizations seeking ARC approval as travel agencies, or to approve locations in the United States where ARC agents may conduct business This corporate accreditation system was set up to approve corporate travel departments in large companies. However, despite the predicted large interest by potential candidates, only a small number actually applied and were accredited.

Another case from the business arena is the request for business accreditation to improve global labor standards and tackle international distress regarding labor abuses (Raynor 1999). This involves a private-sector accreditation based upon the traditional accreditation system in academia, and it has been proposed as a potential approach to help solve labor concerns. Self-chosen accreditation of corporations would be a recurrent system that evolves into a method to ensure that specific and proper labor standards are being met by businesses. This voluntary system has the potential promise to address concerns because the accreditation would be international, it could take action more quickly than other public guidelines or strategies, and it could be used as a promotional tool for businesses today. However, in reality, the true overall motivation to companies is the idea that accreditation is not primarily a means to quality improvement but is really a means of making the business more profitable via external market approval for sales generation. Colleges and universities have a more unclear meaning of what their triumphant results are in regard to their stated goals and, in particular, in regard to the general public, which is their potential customer base—parents and prospective students who are considering enrolling in higher education. In this regard, academic accreditation is important for public assurance and trust, as well as the more recently instituted and evolved goals of enhancement and revitalization of the educational organizations. However, there are still many concerns involving public perception and facts about the accreditation system in America that need to be tackled.

CHAPTER 4

Accreditation under Fire

The system of educational accreditation has been publicly critiqued by those in higher education within accredited institutions and individuals and organizations in the educational community, as well as by a variety of people and organizations outside academe who disparage the usefulness and efficacy of the current structure. Being that the nature of public discourse usually involves complexities and speculation, the truth is that several of these alarms are probably warranted while some are not. Even as early as 1939, Samuel P. Capen, the chancellor of the University of Buffalo, stated that educational leaders around America were tired and distressed at having the educational and financial procedures of their organizations commandeered by careless strangers who had selfish or unclear reasons as their true goals (Capen 1939; Young et al. 1983). Within the accreditation community, some people have described the procedure as intangible, imprecise, shallow, and full of vague meanings that vary among organizations and stakeholders. This is due, in part, to the lack of understanding about the true nature and role of accreditation among the general public or even among the educational organizations that this oversight system serves (Young et al. 1983). While some of the critiques have come from ignorance or misunderstandings about the proper role and power of accreditation, other pessimistic observations are founded on the conviction that the procedure could accomplish far greater outcomes if only certain changes in the accreditation objectives and methods were implemented. These perceptions are reinforced by the constant demands that educational leaders are flooded with to reduce costs and improve professorial and personnel output while continuing to attain additional specialized and professional accreditations for their institutions. Doerr (1983) wrote that "if accreditation isn't slowed as a phenomenon, institutions of higher education may well declare that process is our most important product" (p. 8). While some may argue that process, deliberation, and dialogue are what those employed in academe truly enjoy most in their profession, when faced with a choice to work on additional accreditation processes or additional discipline-related discussion and debate, the faculty and leaders would undoubtedly choose the second alternative.

The critiques of college accreditation these days also include the high cost for the institutions, the burdensome makeup, the supposed superficiality perceived by many colleges and universities, the inequality of the various standards, and the belief that accreditation is for the most part merely a convenient mechanism to achieve certain ends (Dill 1998; Ewell 1994, 2001). Many of the very experienced leaders in academe, including institution presidents, have been particularly vociferous in expressing their disapproval of college and university accreditation procedures. These criticisms may have recently been directed more at the professional accreditation agencies, such as those for business, education, and law, rather than at the primary regional accrediting agencies. This may be due to the positive changes that have occurred in many regional accrediting agencies, from measuring inputs to assessing outcomes and providing practical help to colleges and universities of different sizes and stages of development. For decades since their founding, the professional accrediting organizations in higher education, such as AACSB International and the National Council for Accreditation of Teacher Education (NCATE), measured institutional quality with more easily quantifiable measurements of resources than more variable gauges of academic value. This increase in specialized accreditation produced a rivalry for budgeted funds in medium and large institutions between the different academic specialties, and nearly disqualified the smaller colleges from even seeking these somewhat expensive accreditations. The processes for specialized (professional) and regional accreditation have developed considerably in recent years, yet they are still seen by some as being too proscribed and too broad in scope to delve deeply into real organizational and educational deficiencies in the institutions.

For example, critics complained for years that AACSB International standards were too formulaic and strict with regard to faculty qualifications relevant to practical business experience, and that overall the standards were too unyielding (Benson 2004; Neelankavil 1994). From the 1920s to the 1950s there was a wide range of quality in business education, and the discipline itself was still seeking a balance between theory and practice, and respect in academe. In 1988, the Porter-McKibbin Report stated that although a large amount of progress had been made in the past three decades (particularly regarding the quality of faculty and students), faculty needed more interaction with the corporate sector and the curriculum lacked integration and sufficient global awareness (Porter and McKibbin 1988). Since the early 1990s, business school accreditation has evolved to minimize the perceived weaknesses and has sought to develop a stronger balance between research and teaching. The accreditation leaders have sought to enable business schools to become true schools of a profession, where stakeholders evenly rate the worth of their dual duties to the higher education and corporate world. The AACSB has again adopted new standards for the accreditation of business schools recently, and these revised standards seek, on the one hand, to broaden faculty qualifications, and on the other, to improve the educational quality of the programs for students.

A number of professional accreditation associations and even some regional accrediting agencies in the United States are sometimes seen as avoiding the most basic principles and expectations of the higher education experience. Specifically, this includes the baccalaureate core courses that are normally taught by members of the arts and science faculty. Other strong pessimistic concerns about specialized accreditation include the high cost in faculty and administrator time and opportunity cost in work hours lost by many important college and university managers who participate in a standard accreditation internal self-evaluation study and external team site visit to other institutions. Some would argue that the internal self-examination and the visits to other institutions are valuable processes in themselves. However, the concerns about the value of an accreditation review are particularly strong at many larger and well-established universities where their own academic standards are thought to be clearly above the threshold of most accrediting bodies (Dill 2000; Ewell 1994).

Some critics also view the accreditation procedures as discouraging improvement and disregarding a college's or university's characteristic objectives and the students it was established to serve (Dill 1998). There have been significant revisions in many accreditation standards to address these issues (and some examples will be shown later in this book) of institutions that not only assessed their quality and improved education in their accreditation self-study process, but were also pioneers in the actual accreditation procedure. Despite these increasingly frequent innovations and significant improvements in accreditation, these positive changes do not necessarily reduce or alter the viewpoint of many critics of higher education that accreditation evaluations and the peer-reviewers who perform the evaluations are often unprepared, subjective, and idealistic in their decisions on, or demands upon, the colleges and universities being examined. This is especially wearisome to the very knowledgeable experts who work in postsecondary educational institutions, who generally have worthy principles of impartiality, preparation, and robust characters that do not accept external criticism very well—particularly if the evaluation was not performed in a method that was seen as methodical, learned, practical, and carried out by reviewers who are valued by those in the colleges and universities under scrutiny. Additionally, some accreditation associations are perceived as intently focused and thoughtlessly motivated by accreditation standards that are not appropriate to the institutions being reviewed (Dill 1998). This condition adds to preexisting viewpoints that the entire accreditation system is defective and in need of enhancement.

Theories of higher education envision the accreditation self-evaluation procedures as a continuous process, but in practice this is rarely accurate (Barker and Smith 1998). Higher education institutions normally just think about, prepare for, and execute their regional and professional/specialized accreditation processes just prior to the time when a mandatory review is approaching. Recently there have been some requests for combining accreditation scheduling into strategic and operational planning, and this book describes several examples of managing accreditation processes effectively through proper planning, funding, and evaluation, which are described later in this book. However, for most colleges and universities today, the accreditation process is largely seen as a required and unpleasant task that is put forth by semigovernmental organizations that are not usually viewed by most people as supportive to the institution. The generally unflattering opinions of many people are most likely due to the procedures of accreditation as well as the implementation of this mechanism in the American system of postsecondary education. For instance, although membership in and approval of accreditation associations are commonly referred to as voluntary (Casey and Harris 1979), in reality accreditation is a necessary qualification for institutions that offer to students federal financial aid, and accreditation is also necessary for a plethora of institutional grants. Therefore, the "voluntary label" seems insincere from the start to many people, as there is, in truth, a large financial incentive or necessity to attain and continue regional, if not professional, accreditation approvals. Accrediting associations can encourage accreditation by educating stakeholders in academe as well as the general public regarding the real value of this sophisticated process of peer evaluation and organizational self-review, the benefits of guidance from reviewers and quasi-government identification, and the genuine prospects for revitalization that effectively facilitated accreditation processes can provide. Correspondingly, college and university administrators are challenged to educate their institutional stakeholders about the rewards of pursuing accreditation, their proper management and amalgamation into the organizational processes, and about the foundation of an institutional knowledge base that is always aware of the requirements of accreditation agencies, the best practices of other institutions, and the delivery of a system that is ongoing and more than just an intermittent external evaluation.

Prominent organizations and respected scholars have also criticized accreditation, as exemplified in a 1995 report that was funded by the Mellon Foundation and published by Columbia University (Graham et al. 1995). This essay condemned accreditation as an arrogant intrusion united with insignificant interest that supports the status quo instead of meaningful self-evaluation by postsecondary institutions. The report recommends improvements in the accreditation processes such as concentrated internal self-examinations that converge on instruction and student learning, and examination of true internal quality-improvement methods at institutions of higher education. In addition, perhaps the most comprehensive, and possibly important, official denunciations of college accreditation published in recent years is a report by the American Council of Trustees and Alumni (ACTA) that was offered and given to U.S. congressional leaders who were deliberating the renewal of the Higher Education Act, which includes oversight of accreditation (Leef and Burris 2002; Morgan 2002). A U.S. House of Representatives subcommittee met to evaluate the function of accreditation in higher education, and lawmakers firmly criticized the U.S. accreditation process. Claims that it fails to guarantee academic worth, is short of accountability, and increases the operational expenses of education for institutions and students were made. The ACTA report even condemned the fundamental basis of accreditation, by questioning the connection of accreditation and U.S. federal aid, and proposed cutting this linkage. Some lawmakers were in favor of this step while others were opposed to acting too quickly to change the fundamental principles and importance of accreditation to the American system of higher education at this time.

In particular, the ACTA report asserts that while accreditation of colleges and universities began as a voluntary system to set appropriate standards and list worthy schools, it has developed into an artificial mark of government-sanctioned approval that is now necessary for all educational institutions that seek to provide federal financial aid to students. The report states that nearly all colleges and universities in the United States have earned regional accreditation, and that instead of helping to guarantee educational worth, the accreditation system now simply confirms that a postsecondary institution has what the accreditation agencies consider as suitable resources and internal processes.

In addition, ACTA states that there are major expenses connected with the accreditation process, including the immediate costs of the accreditation procedure and opportunity costs for the lost faculty and administrative leaders' time that educational institutions experience while engaging in the acts of accreditation endorsement. Some lawmakers, many of whom have political objectives in mind with regard to the oversight of higher education policy and governance, have welcomed this report.

Because of these criticisms and questions voiced by other groups and renowned leaders, the regional and professional accreditation associations are making a variety of changes. These modifications include developments in accreditation standards, enhanced linkage to the strategic mission of postsecondary schools, improved internal and external evaluation procedures, generally higher expectations, more stringent membership qualifications, better timelines for reviews, and other improvements. One important topic that is now undergoing examination and analysis is the requirement of scrutinizing institutional procedures for assessing student learning. The regional associations and many professional (specialized) accrediting organizations have been examining methods for educational institutions to properly assess student-learning outcomes for colleges and universities programs, developing plans to employ these measures, and show that the procedures at the institutions which develop the learning outcomes are ongoing. The scope of this exercise in the field of accreditation and the colleges and universities they evaluate has been called extraordinary (Ewell 1998).

There have also been concerns about the status of high default rates on student loans at some proprietary schools, and therefore it has been suggested that the USDE in the U.S government should focus more on reviewing accrediting agencies that evaluate these proprietary schools (Martin 1994). Even in the face of these efforts at changes, and even as many of the improvements are tackling important problems voiced by the accreditation association membership and the general public, there are still many critical problems that are yet to be examined properly. These concerns include increasing the ease of use of public data about college value and concentrating on the nondegree and for-profit providers (Eaton 2001a). In the past, accreditation associations have supplied only a small portion of available data to the general public about the institutional value of the colleges and universities being examined in their periodic accreditation reviews. The justification for this policy may have been genuinely founded on the desire for faith and honesty among peer institutions that were experiencing evaluations. However, growing demands for answerability, government inspection, and escalating tuition prices have produced higher pressure for more candor. Additionally, the ever-increasing visibility of rivalry between traditional colleges and university education providers and proprietary (for-profit) corporations has expanded the demand for increased oversight of the accreditation approval procedures, their importance, and its worth to the people.

The current turning point is therefore based partially on these issues, and partially on recent federal government efforts to change the accreditation system: specifically, a federal bill to reauthorize the Higher Education Act, H.R. 507 (see figure 1). This was foreshadowed in the 2002 "No Child Left Behind Act" that demanded greater accountability and performance from elementary and secondary schools (Farrell 2003). The enactment of this law created many changes in the educational institutions that supplied higher education with students. Subsequently, in the following

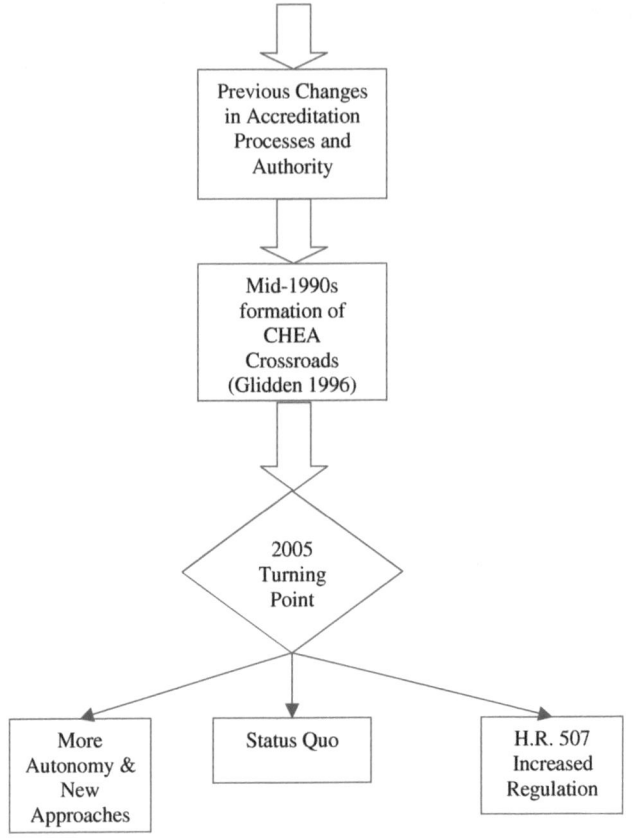

Figure 1 Accreditation Turning Point in 2005

legislative years, the renewal of the Higher Education Act proposed substantive changes to the college-accreditation system. Specifically, legislators sought to address criticisms regarding the supposed problems with the current accreditation system, particularly the pressure to standardize accreditation across the country, increase standards, address changing technology, track transfer students, and generally provide more information than a "yes" or "no" decision to the public regarding the quality of colleges and universities.

Other options for the way in which the evolution of accreditation may turn include maintaining the current system (status quo) of current self-governance with moderate federal oversight and coordination, or perhaps a less-likely shift toward more autonomy by institutions of higher education with new approaches to self-regulation. There are those in higher education who believe that the federal government does have a role in shaping the direction of higher education, but that it should not be intrusive. Three areas where additional oversight and guidance seem to be needed are in lowering financial barriers to access, addressing public and political pressures for improved accountability, and controlling cost increases because the tuition price curve increases are perceived as sustainable in the long-term (Lingenfelter and Lenth 2005).

The new developments and increasing expectations of higher education are provoking the latest reassessment of the roles of the federal government, state education departments, and institutions, and the issue of tuition costs and tax money support for public institutions has drawn in extra scrutiny. The budgetary pressures combined with the accountability issues raise some legitimate issues for consideration and possible changes, but a strengthening of the current system instead of complete realignment away from a long history of success would be a wiser course of action in the area of accreditation.

The CHEA and other associations oppose standardized measures and other aspects of the new legislation (CHEA 2004a, p. 733). The accreditation provisions include a strong emphasis on student learning outcomes, regulation of distance education, transfer credits, and the reporting of public information. The regional associations are against standardized measures partially because of the valued differences between institution types, including community colleges and research universities. The reauthorization bill has also created strong reactions because many of the national associations believe it moves away from the proper balance of voluntary accreditation and government controls over higher education (CHEA 2004a, 2005b). Transferring college credits would be directed by law and not by the colleges and universities, as many believe they should be. The same would be true for distance education, the publication of written learning objectives, and the new reporting requirements that are viewed as excessive, redundant, and costly. Legislative leaders have attempted to fight back against the reaction of the education associations and have sent letters to 6,700 college and school presidents criticizing the lobby groups, who complain that the new bill would not help financially needy students, but instead increase federal intervention in their college affairs and increase the cost of reporting requirements.

In addition, the overall approach in the reauthorization of the 1965 Higher Education Act seems to be that the public does not understand how accreditation works, that there is a plethora of accrediting agencies that increase the complexity for understanding and adherence, and that a much more comprehensive and authoritative program of education and backing is needed to clarify the information available and evaluate how effective the accreditation process is (Glidden 2003). Those not familiar with the broad issues in higher education admissions access in the United States (when compared with those in other countries), differences in student learning styles and institution missions (ranging from community colleges to research universities), and the inappropriate usage of easily quantifiable statistics (such as admissions data, funding rates, library volumes, and student graduate rates) can be persuaded quite easily to strive to change and promote these supposedly important issues, to the detriment of more meaningful determinants of institutional quality today. The use of measures such as graduation rates has recently come to the forefront in some circles, and colleges have been urged to remain open to tracking student progress (Astin 2004; Burd 2005). A UCLA study of 262 institutions found that student graduation rate comparisons should be based on characteristics of entering freshmen, not just the stated rate. The study reported that two-thirds of the interinstitutional variation in baccalaureate-degree completion rates is attributable to differences in entering freshmen. Therefore, accreditation of colleges and universities and attempts to assess or penalize institutions because of their graduation rates may not be appropriate in light of the egalitarian approach in the movement over the past century from elite to mass to universal higher education (Trow 1973).

At the time of the writing of this book, the latest incarnation of the Higher Education Act seems to have slightly tempered the increase in regulation and put some of the more contentious issues for accrediting agencies and higher education institutions aside for now (Bollag 2005). The issue of public disclosure has changed from requiring the release of an accrediting agency's findings on institutions to only requiring such disclosure when an institution is put on probation. This is an attempt by the legislators to create a stronger warning to colleges and universities to improve their deficiencies, but does not require the accrediting agency to withdraw accreditation. Another element of the proposed changes that was seen as an increase in government interference is the forced acceptance of transfer credits from one institution to another. This academic decision is seen as sacred by many institutions, and the latest House and Senate bills have mollified the proposed law to only direct colleges and universities to not reject credits earned elsewhere just because the original institution is not accredited by one of the six regional accrediting agencies in the United States. Many of the for-profit institutions have been seeking to have their college credits more easily transferable to more academically oriented institutions, but since these institutions are normally only accredited by some of the specialized national accreditors, such as the Accrediting Council for Independent Colleges and Schools and the Accrediting Commission of Career Schools and Colleges of Technology (see complete list in Appendix A), this was frequently not acceptable. The latest regional association requirement by legislators includes a somewhat milder change, yet is still an increase in oversight and can be seen as part of the turning point in higher education accreditation.

In addition, other recent developments in the debate regarding the institutional appeal process for accreditation decisions and the power of states to function as accreditors are still moving forward in this changing era of accreditation oversight (Bollag 2005). The legislation that is nearing approval contains the right to appeal an accreditation agency's decision to place the institution on probation, whereas the current policy only allows appeals for denial or revoking of accreditation. The accrediting associations oppose this change because they believe that probation does not truly represent a change of status for the institution and serves as a strong warning to improve. They believe that this change will slow down the appeals process and make the use of probations less likely, resulting in less-effective quality assurance. The other contentious issue is the power of states to function as accreditors for colleges and universities, which the government hopes will increase competition in the accreditation system. However, higher education groups see this development as a break with the long history of successful peer-review in the United States, one that allows state bureaucrats to increase regulatory oversight. In addition to political intrusion into academe, there is the potential problem of questionable institutions then having the option to persistently pursue accreditation in a multitude of venues to achieve a license to operate.

Related to this potential problem of questionable institutions seeking accreditation, there has also been a series of recent concerns regarding diploma mills and accreditation. While established and reputable institutions normally seek regional, specialized, or national accreditation, there are other supposed agencies offering approval that are not widely respected. (See Appendix B for a list of accrediting agencies that are not recognized under the Generally Accepted Accrediting Practices (GAAP), and not recognized by either the Council on Higher Education Accreditation, the USDE, UNESCO, or by education departments or ministries of major countries.) The United States government refers to Webster's definition (on the www.ed.gov website) of a diploma mill as "an institution of higher education operating without supervision of a state or proficiency agency and granting diplomas which are either fraudulent or because of the lack of proper standards worthless" (Webster 2002). Diploma mills frequently proclaim accreditation by a false or phony accreditation agency to invite a large number of students to their collegiate-level degree programs and make them appear more genuine (USDE 2006, p. 898). Since these diploma mills are not accredited by an accepted bureau, these supposed institutions of higher education will not be found on the list of approved colleges on the USDE's List of Nationally Recognized Accrediting Agencies (USDE 2006) or the CHEA's (CHEA 2002 p. 593) list of participating or recognized organizations. It is suggested that institutions and individuals check that the institution being considered is accredited by a nationally recognized agency, and that those accrediting agencies recognized by the USDE are approved primarily for the purpose of obtaining federal student loan funds. False accrediting agencies are often just for show or nefarious purposes (Bartlett 2004; Bartlett and Smallwood 2004), and commonly offer accreditation to institutions for a monetary fee and without a thorough examination of the college or university's programs or professors. These false accreditation agencies may use organizational names that are comparable to established and well-known accreditation associations, and even add genuine higher education institutions into their list of officially accredited member institutions. Colleges and universities should have an effective accreditation management strategy to oversee all accreditation issues and prevent potential problems. Criteria for official recognition as an accreditation agency are listed in Appendix E.

CHAPTER 5

The Need for Accreditation Management

Despite the many criticisms about accreditation of higher education at this time, institutions and the general public still have a fundamental desire and true need for identification and revitalization of postsecondary institutions by societal administrations such as the federal government. Twenty years ago, a research study surveyed 520 institutions accredited by the American Council on Education in 1986 and discovered that 90 percent of participants surveyed concurred that education accreditation provides a practical directory of institutional value and worth (Andersen 1987). The study also found that a preponderance of thought that accreditation is perceived as a practical instrument for self-evaluation and a incentive for organizational improvement. Nevertheless, only 20 percent of the survey respondents reported it to be very useful as a measure for entitlement to offer federal money to students. This perception brings up an inquiry with regard to the potential divergence between the summative and formative objectives of the accreditation procedures. This potential conflict may be supported by established management theories related to the separation of such goals in performance appraisals (Meyer et al. 1965). The main idea is that appraisal of functioning should not be connected directly with development strategies, in order to augment the prospective benefits of each educational objective. Can external collegiate colleague-evaluators in an accreditation review offer genuine and comprehensive guidance for improvement when the process also requires that the evaluators give a formal decision on institutional or specialized program quality? Some people who are uncritical or perhaps unaware of this perspective would probably say that there is no difficulty for accreditation representatives to have dual roles as both reviewer and advisor. However, others might think that there should be more of a division of these roles to help guarantee the greatest potential quality of the accreditation evaluations and suggestions for enhancement. Despite the cause for the varied views by educational members about the effectiveness of granting the accreditation system as the arbiter for permission to grant Federal student financial aid and obtain federal funds, the established arrangement will probably remain for many years.

Colleges and universities should contemplate both the summative and formative objectives of the accreditation process while designing and implementing strategies that are appropriate for planning and management of accreditation to accomplish the highest potential outcomes from both of these goals.

The accreditation process has been described as a "voluntary process conducted by peers via non-governmental agencies to accomplish at least three things: . . . to attempt on a periodic basis to hold one-another accountable, to achieve stated appropriate institutional or program goals . . . to assess the extent to which the institution or program meets established standards" (Kells 1994). Consequently, based on the abovementioned thinking, it can be promulgated here that accreditation management, from the viewpoint of the college and university leaders, is required to help guarantee ongoing organizational development and government identification through efficient management functions such as planning, budgeting, organizing, and controlling the required regional and desirable specialized accreditation processes in an orchestrated endeavor that satisfies several functions. Accreditation management means evolving beyond the perception that accreditation reviews are merely a periodic institutional irritation by educational stakeholders that needs to done with nominal energy and conformity. Successful facilitation and administration of the various specialized and regional accreditation processes means using the considerable information, external power, and fundamental scholastic energy available for enacting important developments in educational institutional effectiveness and student-learning outcomes. Lately, assurance of student learning has become a very important measure for regional and specialized accreditation standards, and colleges and universities must strongly concentrate their operational processes, funding, and strategic planning on implementing this as one of their most significant objectives in their plans. With the exception of a few primarily research-based postsecondary institutions, assurance of student learning and the mechanisms therein have been determined to be the primary institutional goals which accreditation is now assessing.

These latest series of developments in regional and specialized accreditation are an encore to other recent improvements consisting of strongly needed changes, of which nearly all have been for the overall good of education. Roughly two decades ago, the regional accrediting agencies in the United States were disconnecting their strategies from the traditional quantitative evaluations that largely involved data input such as budgets and the number of books in libraries, and evolving toward a more qualitative examination which looked at broader organizational and educational matters (Young et al. 1983). Because the traditional approach encouraged conformity of institutions, critics rightly pointed out that college and university individuality was being neglected and even penalized. American postsecondary institutions, and even individuals in the uniquely independent culture, saw conformity as a weakness that needed to be dealt with and hopefully reversed. Interested readers and scholars of higher education know that the origination of college accreditation had as one of its primary goals to ensure minimal quality, can see how it is the de facto standard of easily and quickly measurable institutional quantitative data which were used for decades as a expected gauge for educational quality. As the educational system has evolved into the complex and diverse range of institutional types, goals, programs, locations, students served, and technologies available, the traditional accreditation system was naturally becoming obsolete. The accreditation process itself was dependent on external review, and then moved toward more reliance on self-evaluation and self-improvement by the educational organizations. The accrediting body moved from merely judging to encouraging and assisting educational institutions to improve their quality.

Recent accreditation reforms in regional postsecondary accreditation have resulted in notable developments in the fundamental and political relationships to the education stakeholders and to their strategies for quality review (Eaton 2001). Accreditation standards have been updated to not only examine quality improvement initiatives, but also to position the regional accreditation procedures and self-evaluations to implement national quality review expectations. Some of these expectations include the development of an increasing number of distance-learning courses at traditional colleges and universities, completely Internet-based distance-learning programs and institutions, the growing need for international quality reviews by U.S. accreditation agencies, the escalating awareness to assurance of learning, and even attaining increased efficiency through synchronization of reviews among various accrediting associations. Aside from the regional accreditation, specialized accreditations, which are largely for professional programs in business, education, medicine, and law, necessitate greater institutional awareness, budgetary support, and senior administration backing. Professional or specialized college program accreditation is viewed by some in the field as a costly, problematic, but unfortunately required effort for quality institutions, which often created either great problems and were a letdown or were seen as a wonderful accomplishment. One college president recently stated that specialized business school accreditation such as by AACSB International, was indeed difficult but it also proposed optimism for advantageous reforms (Dill 1998). Many colleges, universities, employers, students and parents now see that this kind of specialized accreditation is very important to push for communal self-evaluation, development, and program enhancement. Specific professional accreditation is beneficial because it not only shows how specialized programs can be enhanced, but also helps colleges and universities to recognize the extra funding that is often required and the additional resources that are frequently needed for superior education in those professional fields with exclusive requirements. Today, there are many institutions that manage the maintenance of regional and several specialized accreditations in manners that are fair to both the non-professional programs as well as the accredited degree programs, for overall institutional excellence, student-learning outcomes, and organizational success that benefits everyone at the institution. Several examples of cooperative inter-accreditation management strategies will also be described later in this book.

It is important to understand the management of specialized accreditation as it is connected to other accreditation procedures, along with the role of educational leaders in implementing change within association and accreditation-sponsored reforms. A research study on the managerial function of professional school deans within postsecondary schools that are executing revised accreditation standards by a specialized accrediting associate, discovered that faculty involvement can be directed to support transformation and accreditation procedure-related improvements, but only to a restricted degree (Henninger 1998). It has been found that academic disciplinary and local institutional limitations greatly influence faculty members and their individual professional expectations. It is understandable how in a highly decentralized multidisciplinary institution such as a college or university, colleagues and influence from other sources may negatively impact professors' willingness to fully collaborate and involve themselves in significant organizational development actions. There is a need for accreditation management strategies to negate this faculty effect through effective planning, internal promotion, budgetary support, and personnel choice that reduce the consequence of these issues. As mentioned earlier, the fundamental nature of higher education institutions today tends to disperse the authority and power among institutional principles, standards for specialized accreditation, and the various academic disciplinary viewpoints that significantly affect faculty member expectations in ways that may restrict the abilities of institutional leaders who are accountable for the accreditation procedures. The faculty management style and limitations must be thought-out when planning for accreditation, inter-accreditation strategic collaboration, and pioneering reforms in pursuing accreditation approval and internal revitalization.

There are significant internal and external pressures for managing the accreditation processes in colleges and universities today (see figure 2). These include the noteworthy requirement for federal funding through U.S. government-backed student loans and grants, the selection among the various specialized accreditation agencies (and certain postsecondary institutions may choose among the regional associations for accreditation), the rising expenses associated with accreditation, the intra-institutional and societal demands to enhance and confirm student learning, the importance of getting practical advice from contemporary institutions, and the interdepartmental competition in colleges and universities for budgetary resources. Frequently, academic departments and even nonacademic areas have the choice to

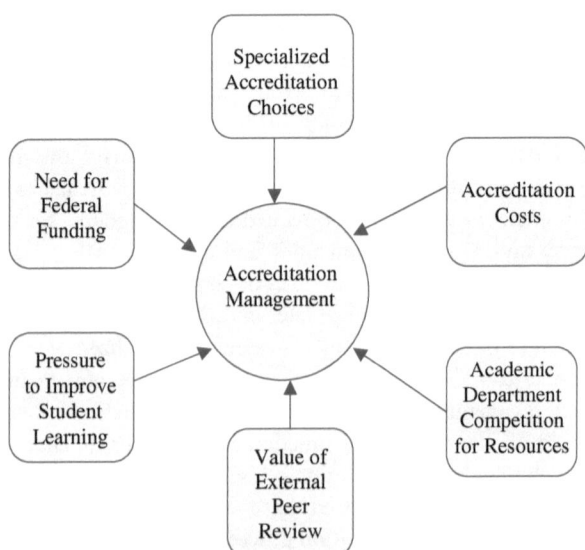

Figure 2 Need for Accreditation Management

work at different specialized accreditations, such as professional divisions/departments, counseling, and athletics. However, institutions must decide, through operational and strategic planning, whether to pay for and otherwise maintain these costly accreditations for all departments which decide to pursue them.

If colleges and universities would apply a managerial approach to accreditation pursuits, then choice would be made based on several important issues such as the institution's mission, multiyear strategic plans, short-term goals, available budget funds and resources, personnel qualifications, and other factors. If these pressures are not addressed directly by colleges and universities, then there is a very strong possibility that outside forces will intervene. The recent hearings in the U.S. Congress on college accreditation as part of the renewal of the Higher Education Act has called for more responsibility by the federal government, using or expanding the accreditation agencies (Morgan 2002). Additionally, many individual state oversight agencies have started to implement statewide standardized testing of college or university graduates, as an intended assessment for student learning which can be linked to state funding approval for institutions based on the student scores. If individually and collectively higher education leaders do not oversee accreditation reforms internally, then it appears that the external societal powers of academe will manage to do it. Some thought leaders in higher education believe that the problem for colleges and universities is not the process of creating more standardized testing, but how institutions can assess internal procedures for performance improvement that are coordinated across all postsecondary schools, and are also convincing to those outside academe (Ewell 2001). It is therefore proposed here that internal management of accreditation processes by colleges and universities themselves will deal with these forces and facilitate ongoing acknowledgment of accreditation by the general public and yield greater internal revitalization.

PART TWO

The Accreditation Process

CHAPTER 6

Getting Started on Accreditation

A large number of postsecondary institutions have developed internal periodic review procedures and they also partake in official regional accreditation procedures along with several optional specialized program or school accreditations (Moore and Diamond 2000). These internal institutional and external peer-review processes are usually alike in their primary procedures, but have largely followed rather than pioneered the developments in higher education. Recently, U.S. accreditation standards have been mainly focused on guiding institutions to be directed by their missions and essentially guide college and university improvements. Most agencies have become deeply involved in facilitating a focus on student learning outcomes or assurance of learning. Therefore, internal institutional processes that develop in coordination with an effective accreditation management policy ought to concentrate on the college's or university's mission and objectives for student learning, along with coordinating the accreditation timing procedure, planning, staffing, and budgeting in conjunction with the internal reviews as well as the vital operational and strategic institutional goals. As this examination of effective accreditation management strategies begins, it is essential to first consider the particulars of a typical accreditation procedure. The specific processes of many regional and specialized accreditation agencies are often rather alike, and usually have familiar steps in their processes.

A reader who has heard the term many times, may still wonder exactly what accreditation means. One book defined accreditation as "a process by which an institution of postsecondary education evaluates its educational activities, in whole or in part, and seeks an independent judgment to confirm that it substantially achieves its objectives and is generally equal in quality to comparable institutions or specialized units" (Young et al. 1983, p. 21). As mentioned previously, accreditation began in the United States as a system of achieving nongovernmental colleague review of postsecondary educational institutions and individual programs. Using private educational associations of regional or national (specialized) scope, the United States has implemented standards that are intended to mirror the behavior of good educational curricula and has created systems for assessing institutions or programs to verify whether colleges and universities are working at fundamental levels of apparent quality. According to the U. S. government, the basic procedure in a typical accreditation process includes:

1. Standards: The accrediting agency, in collaboration with educational institutions, establishes standards.
2. Self-study: The institution or program seeking accreditation prepares an in-depth self-evaluation study that measures its performance against the standards established by the accrediting agency.
3. On-site evaluation: A team selected by the accrediting agency visits the institution or program to determine firsthand whether the applicant meets the established standards.
4. Publication: Upon being satisfied that the applicant meets its standards, the accrediting agency grants accreditation or pre-accreditation status and lists the institution or program in an official publication with other similarly accredited or pre-accredited institutions or programs.
5. Monitoring: The accrediting agency monitors each accredited institution or program throughout the period of accreditation granted to verify that it continues to meet the agency's standards.
6. Reevaluation: The accrediting agency periodically reevaluates each institution or program that it lists to ascertain whether continuation of its accredited or pre-accredited status is warranted. (Government 2002)

The most important steps in this sequence for individuals at most colleges and universities are steps 2 and 3, because they are vital to the procedure and need significant professional management. At the institutions, attention is usually concentrated on all approaching self-study procedures (also called self-evaluations) and the impending on-campus visit by a peer-review team of academics from other accredited colleges and universities who are selected by the accrediting agency. But effective accreditation managers should be aware that these accreditation steps are only a part of a total accreditation management strategy described in this book for colleges and universities that wish to organize, harmonize, and capitalize on the value of the assorted accrediting procedures that most institutions normally undergo at recurring intervals. Even so, more colleges and universities are now using the self-study process itself as a central mechanism for improvement. For instance, the institution's leadership at the University of Vermont joined forces with faculty members, staff, and students to leverage the accreditation process as a vehicle, or "chariot," for institutional makeover (Martin et al. 2001). The accreditation self-evaluation step was both the motivation for change and the basis for objectives that brought together the academic departments in discovering an institutional mission. In this example, the accreditation self-study procedure was more than just an opportunity to exemplify continuous improvement at the university. The procedure was more than a reaction to standard accreditation expectations and the required need for strategic planning than a proactive management method. In many colleges and universities, this method may be essential and helpful to implement changes when there is intense faculty and administrative antagonism, apathy, or other issues that often hamper strategic planning.

Since there are a large variety of management models, tactics to solve problems, and personal management techniques among college administrators and faculty to examine, there are also numerous probable obstacles to an effective self-study (Kells 1994). In fact, these obstacles are not exclusive to postsecondary schools, and are relatively widespread in many elementary/secondary schools and nonprofit, service-oriented organizations. This is especially true of those that are publicly funded, such as government bureaus and research institutes with a large number of highly educated professional employees (Goodstein 1978). The prospective obstacles to self-study that can arise in these types of organizations include:

- Unclear goals: The goals of these institutions are very complex and difficult to clarify. It is often difficult to reach agreement among the highly educated professionals involved, and the nature of the evaluation process means that there are few easily measured outcomes to demonstrate achievement of goals.
- Lack of readily available and useful data: In recent years, educational institutions and accrediting associations have been attempting, with mixed results, to standardize and agree upon what data definitions are needed, and how the data should be examined and reported.
- Organization and governance: Educational institutions often have a very flat organizational structure with decentralized administrations that share and delegate their powers substantially. Colleges and universities often have faculty governance bodies, administration structures, unions, or some combination of these that can inhibit or complicate self-evaluation and changes. In addition, the tenure process has given many faculty and administrators the perception that faculty are, in effect, volunteers that need to be dealt with and managed in a very delicate manner because of their supposed antimanagement views.
- Staffing issues: The professionals in higher education, both faculty and administration, are usually very specialized in areas that may be unrelated to areas of their responsibility with regard to a self-evaluation study. A person who is a good teacher in a particular academic discipline may not have any training in running a department or much-needed management skills, and therefore this may hinder the self-study process.
- Interaction among groups: Inter- and intragroup dynamics may also be a problem. In addition to their individual specialties, academics often tend to examine all problems from a disciplinary or departmental perspective, regardless of whether the issue under consideration needs a broad institutionwide examination and solution.

There are also other hindrances to self-studies at educational institutions, such as the academic nature of in-depth problem analysis for all tendencies for conflict-avoidance, histories of poor planning procedures, and a general dislike of direction or guidance from highly educated independent thinkers who are often too proud to accept other opinions. Regardless of these potential obstacles, self-evaluation studies in accreditation processes can be, and usually are, beneficial to the institution. Many accreditation associations have been making modifications in their processes even though they too encounter significant internal institutional obstacles at colleges and universities, as well as rising external disparagement.

A study by the Council for Higher Education Accreditation (CHEA) of 81 regional and specialized accrediting agencies discovered that many of the associations are diverging from conventional accreditation methods which are more individualistic collaborative programs (Safman 1998). These innovative strategies may help develop how self-studies are managed, how visiting teams are shaped, how the visitation team reports are written, and how institutional information is collected and shared among accreditation associations. Cooperative accreditation approaches that can be part of an effective accreditation management strategy will be examined in a later chapter of this book. Many regional and specialized accrediting associations have been assembling collaborative institutional evaluations by mutually assigning a joint review team that collects data from the colleges or university, interviews the institution's faculty and administration, assesses facilities, and authors a single accreditation report for both accreditation associations. In order to accomplish this cooperative evaluation scheme, the institution agencies must require that both accrediting associations agree on a review calendar, visitation team setup, and the topics of the self-study. Nevertheless, the accrediting organizations do preserve the authorization for their individual accreditation decisions on the particular institutional assessment with regard to their standards. College and university accreditation managers should be informed that these cooperative self-evaluations do provide some advantages to the institution being reviewed and may be viewed by some accreditation critics as a temporary solution to the perceived coinciding, repetitive, and unnecessary costs associated with the various accreditations. There are other important matters about accreditation processes that also need to be addressed, such as overall responsibility, diligence, and the likely conflict involving the formative and summative purposes. In spite of this, postsecondary institutions need to encounter and properly address the current accreditation system as it is; therefore, to achieve successful recognition and renewal, effective management of college and university resources is strongly required.

There have also been other advances in methods of accreditation procedures, such as three-year cooperative accreditation reviews, collaboration with international universities and associations, alliances of various disciplinary subspecialties, options for self-study topics, decreased number of campus visits, condensed visitation team size, and the requirement for accreditation members to complete annual questionnaires to collect institutional data. Accrediting agencies can share institutional information and create common reporting layouts using the multiyear approach because it focuses the accreditation reviews for the institution within a three-year time frame. In regard to international accreditation collaboration, this approach usually involves visiting teams from the United States who link with accrediting representatives from other countries to carry out collaborative reviews for various institutions. In the coming years this may assist the overall comprehension of the variation in the accrediting and quality assurance issues in an increasingly global economy with increasing internationalization of education organizations. Various academic and professional subspecialties such as allied health and related subjects are now looking to cultivate common expert standards of good practice that can also diminish costs for institutions and decrease review redundancy. A significant enhancement that has become popular is the availability of accreditation issue focus choice by institutions in the self-evaluation process, particularly for larger colleges and universities that hope to highlight and improve specific managerial troubles they may be experiencing, such as fund-raising, organizational governance, strategic or operational planning, or other significant issues. Accreditation management procedures should clearly reflect the benefits leveraging all accreditation self-study processes for these reasons as part of their total strategy. Additionally, relatively recent accreditation management strategies such as reducing the number of accreditation review team visits and the number of visiting team members can also be planned by institutions to decrease costs and improve effectiveness. However, one development that may increase institutional accreditation costs and potentially cause issue definition problems is the requirement of annual questionnaires by different accreditation associations. At least one specialized accrediting association (AACSB International) now requires all college and university members to complete lengthy data surveys online each year, which appropriates administration or faculty time and resources for uncertain returns. Additionally, accreditation association questionnaires with sizeable quantities of numerical data may be viewed by some people as a regression back to the old quantifiable input measures that accreditation agencies started with many years ago, instead of a progression toward innovation and enhanced quality improvement strategies through effective external recognition joined with valuable internal institutional renewal.

Academic professional associations, government-approved agencies, and other organizations that administer postsecondary accreditation programs have traditionally been advised to arrange specific restrictions and objectives. Consequently it is important for college and university accreditation managers to recognize what the accrediting agencies are seeking, understand the importance and costs of the specific accreditation(s) being considered, and what kind of management is required for attaining the highest return on the institutional investment in seeking the accreditation(s). When the accreditation associations begin to develop standards and policies for educational institutions, they usually seek to identify the requirements needed for accreditation approval from the educational market itself (Hamm 1997). The various vision and mission statements of colleges and universities are also studied, along with the configuration alternatives for administration of the accreditation system. The specific accreditation standards are then finalized, leadership and authority is selected, procedural issues are delineated, publicity and support is then promoted to prospective institutional members, and initial applications for accreditation are then received. Postsecondary educational institutions must then determine which accreditation, or accreditations, correspond with their vision, mission, and strategic goals. Colleges, universities, or programs therein may then start the process of pursing initial accreditation or reapproval for continuing to be accredited by the association(s). An example of accreditation eligibility requirements for the largest regional accreditation agency in the United States (the North Central Association of Colleges [NCA]and Schools—the Higher Learning Commission) is listed in Appendix C.

CHAPTER 7

The Application for Accreditation

Educational institutions seeking initial accreditation must first establish their institution's commitment to pursuing the accreditation, determine if they are eligible, and then formally apply. To determine the institutional commitment, it must be emphasized to the educational community applying for accreditation that the effectiveness of self-regulatory accreditation depends largely upon an institution's acceptance of certain responsibilities, including involvement in and commitment to the accreditation process (NASC 2002). An institution is expected to apply and then conduct a self-study within the interval specified by the accrediting body. At the conclusion of the self-study, the educational institution must be willing to accept an honest and forthright peer assessment of institutional strengths and weaknesses. For regional accreditation, the self-study will assess every aspect of the institution; involve personnel from all segments of the institution, including faculty, staff, students, administration, and the governing board; and provide a comprehensive analysis of the institution, identifying its strengths and weaknesses.

An institution must be committed to participation in the activities and decisions of the agency. This commitment includes a willingness to participate in the decision-making processes of the commission and adherence to all policies and procedures, including those for reporting changes within the institution. Only if institutions accept seriously the responsibilities of membership will the validity and vitality of the accreditation process be ensured.

Because institutions of higher education are usually committed to the search for knowledge and its dissemination, integrity in the pursuit of knowledge is expected to govern the total environment of an institution. Each member institution is responsible for ensuring integrity in all operations dealing with its constituencies, in its relations with other member institutions, and in its accreditation activities with the regional accrediting association. Applicants for accreditation are expected to provide access to all parts of their operation and to provide accurate information about the institution's affairs, including reports of other accrediting, licensing, and auditing agencies. In the spirit of collegiality, institutions are expected to cooperate fully during all aspects of the process of evaluation: the preliminary visit in preparations for an evaluation visit, the evaluation itself, and any follow-up to the evaluation visit. Institutions are also expected to provide the accreditors with information requested during evaluations, enabling evaluators to perform their duties with efficiency and effectiveness. Once these expectations are accepted and understood, then it must be determined whether the institution is eligible before the formal application can proceed.

The regional accrediting bodies in the United States normally accept applications from educational institutions that are concerned predominantly with secondary or higher education; have characteristics commonly associated with higher education; and meet the eligibility requirements. For higher education, the principal programs of eligible institutions will be degree related and will be built upon knowledge and competencies normally obtained by students through a completed high school or secondary school program, or its equivalent. Such programs will be based on verifiable knowledge that has been subjected to examination by competent academic persons and by established practitioners of the arts, sciences, crafts, and professions. Although diversity of requirements is expected among candidate and member institutions, the course and degree requirements of an applicant institution must also be congruent with those of the broader higher education community that the regional accrediting agency represents.

Eligible institutions may properly offer programs that the accrediting body would not define as higher learning (e.g., introductory courses in subjects that some students may have missed in high school, and courses and special programs specifically constructed to assist students to be successful with college-level coursework), but these are offered in addition to the courses and programs relevant to their mission.

The characteristics of an educational institution and the conditions required by a typical regional accrediting association as a candidate for accreditation, for initial accreditation, and for continued membership are listed in Appendix C. Each characteristic or eligibility requirement is an expected level of performance or precondition that relates to the appropriate standard shown in parenthesis. These are essential eligibility requirements that must be met for consideration of candidacy for accreditation status. Once the educational institution has determined that the eligibility requirements are met, the institution can pursue its candidacy for accreditation status. This offers developing postsecondary institutions the opportunity to establish a formal, publicly recognized relationship with the regional accrediting organization. It should be noted that operating nonaccredited higher institutions that meet the basic eligibility requirements may normally apply.

The candidate for accreditation status is an affiliated institution with a nonaccredited relationship with the commission or accreditation agency. Only accredited institutions are normally members of the accrediting association. Candidacy indicates that an institution is progressing toward accreditation, and attainment of the affiliate status does not ensure accreditation. Attainment of candidacy status is the usual outcome of approval by the regional accrediting body after two separate, sequential stages. Usually these steps are the application for Consideration and a Self-Study and Evaluation Committee Visit for Candidacy (NASC 2002).

The actual application for consideration requires that the chief administrative officer and governing board of the institution state that they believe the eligibility requirements are met with a letter of application signed by the chief administrator that is submitted to the accrediting body. A fee is also charged. The regional accrediting and specialized accrediting bodies normally charge a fee of approximately $1,000 to $5,000 or more for the initial application for candidacy. Other requirements include a thorough written response to each of the eligibility requirements; plans for institutional development; a current catalog; a current budget and audited financial statement; articles of incorporation and bylaws (or charter, if the institution is independent); and, when appropriate, proof of state authority to grant degrees. Institutions that are in candidacy may also be appointed candidacy advisors from an accredited institution (which may add additional costs for candidacy) or be encouraged to hire consultants to evaluate their organization for concerns to address during the candidacy period.

If the accrediting body judges that the institution appears to meet the conditions of eligibility, the chief administrator of the institution is advised subsequently to proceed with the institutional self-study for candidacy, and tentative dates for an evaluation committee on-site visit are set. However, specialized accrediting associations such as AACSB International normally require business schools to participate in a five-year candidacy process. If a regional accrediting commission instead determines that additional information is needed, it may request the institution to host a visit by a representative from the accrediting association. Some other specialized or regional accrediting agency may have a shorter candidacy period.

CHAPTER 8

The Self-Evaluation Analysis

The second step in the accreditation process is a very important and detailed institutional or individual academic program self-evaluation that is performed by the college or university following the requirements of the accreditation agency whose approval is being sought. This self-evaluation is the primary means that accreditation associations currently utilize to determine quality and to aid postsecondary educational institutions' development. A basis for strategic planning at the institution can be provided by the self-study process and may lead to enduring institutional organizational research and self-analysis that enhances institutional openness and provides a basis for employee development (Young et al. 1983). In managing the accreditation process, institutional leaders should strive to make the process as effective as possible by positioning the self-study as internally motivated rather than just as externally focused pressure from the accreditation association (Jones and Schendel 2000). Complete commitment to the accreditation self-evaluation process should be expressed publicly to the educational community by the institution's leaders (Rieves 1999). This will help motivate the faculty, students, and midlevel administrators to participate more fully in the accreditation process and thereby amplify the chance for significant institutional improvement. In a later chapter, examples of innovative approaches to plan the self-study as part of an overall accreditation management strategy for colleges and universities will be illustrated. At this point it is important to emphasize that the self-study should be planned by the senior administrators with the objective not only of earning accreditation, but also of accomplishing specific institutional goals that would benefit everyone involved. The process of conducting the accreditation self-study can be utilized as an informed procedure to elucidate institutional objectives and measure goal accomplishment. For instance, Michigan Tech decided to systematize its regional accreditation self-study based on eight strategic goals, and in the late summer to early fall of the first year in their self-study process they undertook five specific actions (Walck 1998). The five actions taken were: selecting the steering committee and subcommittees; creating written committee charges to guide the committee work; developing a SWOT (strengths, weaknesses, opportunities, threats) analysis to focus the committees and departments on evaluation; conducting an orientation to distribute information and create buy-in; and publicizing the self-study process widely. Although developed as a business planning technique, a SWOT analysis can be a useful strategic planning tool in higher education that invites planners to assess the institution in an informed manner from different perspectives, and can neatly summarize the results of different accreditation subcommittee reports that are generated in a typical self-study procedure. The generalized SWOT analysis that was authorized for use by Michigan Tech is shown in figure 3.

Part of the SWOT analysis involves studying benchmark institutions, and a list of suitable comparable colleges or universities can be proposed and selected by self-study committee members. This selection of benchmark institutions can be decided using colleague or peer-type institutions, aspirant institutions that are striving to become equal, mission-based, student enrollment size, and student market. The specific distinctive institutional competencies and competitive shortcomings can be revealed using informal primary and secondary research sources or through formal structured benchmarking programs. Government agencies, academic associations, and accrediting organizations such as AACSB International sometimes offer statistical services to compare institutions or programs through annual survey information collected from its Business School questionnaire. Other third-party benchmarking information providers include Educational Benchmarking Incorporated (EBI), which offers Stakeholder Assessment Studies, and the Consortium for Benchmarking for Higher Education Benchmarking Analysis (CHEBA), which provides a forum for the exchange of performance measurements. Benchmarking in higher education has become a widely used management technique to gather comparative data, learn best practices, and adapt improvements to home colleges and universities that conduct these studies (Alstete 1996). This valuable tool can be used as a data-driven technique to support an accreditation self-study process and can often reveal the strengths of institutions to the external reviewers, as well as any weaknesses that may need institutional plans for improvement.

The self-evaluation process for regional accreditation is usually the responsibility of the chief academic officer at colleges and universities, whose position is often titled vice president for academic affairs or provost. For professional school or specialized accreditation self-studies, the responsible person in the academic department or school is typically the department chair or dean. Other steps in the self-study process to plan for include the choice of institutional faculty and administration members on the official self-evaluation committee, the related intragroup dynamics, responsibilities for report writing (Young et al. 1983), and the use of virtual team technology. As in the selection of committee members for accreditation on-campus visits, it is also crucial to properly select a representative, suitable, and operation-minded institutional self-study team from a mixture of academic departments in the college or university community (Bartelt and Mishler 1996). Self-study committee team members who are chosen should typically be experienced academics in the appropriate department, with excellent oral and written communication abilities, and have the skills necessary to work collaboratively on institutionwide committees

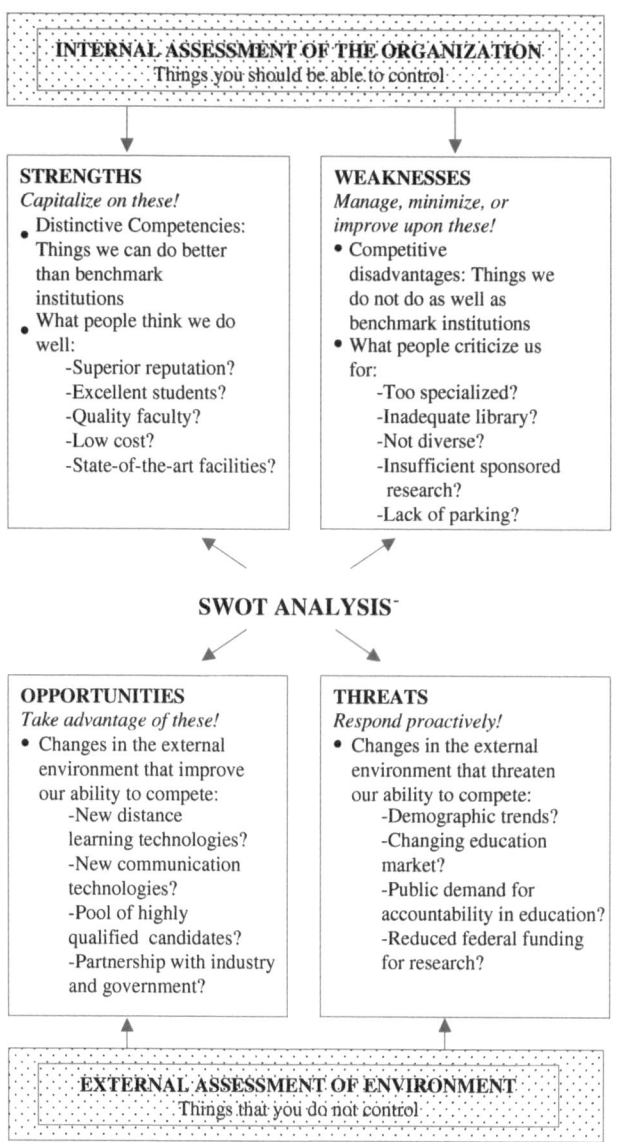

Figure 3 Self-Study SWOT Analysis (Walck, "Organizing and Selling the Self-Study Process," 235, used with permission)

such as the self-evaluation team. The central intragroup dynamics to think about include the classic organizational behavior stages of group development and development of group leadership on the self-study committee. The concluding outcome that the self-study team will produce is a formal written self-evaluation report; consequently exceptional writing abilities are strongly needed and should be confirmed by the senior administration by examination of previous writings prior to selection and appointment to the self-study team. One attractive and fairly recent technique to consider using in the self-study report-writing procedure is the availability of virtual team technology that uses modern Internet-based information e-learning systems for committee interaction, research, operations management, and producing written document drafts. It has been demonstrated that committee or team performance can be improved using these e-learning systems, which are already used for course instruction at many institutions and have many useful qualities that a self-evaluation team can make use of (Alstete 2001). These qualities include team-member discussion boards that can ease intrateam communication among members, digital drop boxes to exchange self-study documents and other material, an announcement web-page for the committee chair to help lead the team, a page for external accreditation Internet links to assist the team explore specific issues, and the capability to have virtual live online chats for real-time interaction, consultation, and instant feedback with team members and others for instructive live discussions. (See the chapter on new strategies for accreditation for more information on using online course software to prepare for an accreditation self-study.)

The common stages in a self-study procedure usually include getting ready for and planning the self-study, organization of the review, carefully observing the process, involving peers in the study, and combining the cycles of study and planning (Young et al. 1983). It is important to set up the leadership and internal rationale for the self-study during the preparation phase, pinpoint a detailed list of college or university needs and topic, and recognize local circumstances to show in the self-study plan. Two fundamentals of a winning self-study that were documented at Pike Peak Community College during their reaffirmation of accreditation were the full backing of the College's president and stakeholder community and the recognition of common issues and organizational direction (Kemling 1994). This shows that a self-study report should not just be a compilation of various committee or accreditation team reports bound together under a common title. A high-quality self-study can bring the members of an educational institution together in search of a common course with resilient leadership and vigorous community involvement. In addition, effective organization of the self-study should also properly describe team members' tasks and roles (this confirms the management structure, which is usually an accreditation steering committee comprising subcommittee or task force chairpeople), choose the best members to familiarize and train, secure needed resources, establish timelines, and set coordination of communication methods. Goncalves (1992) states that it is extremely critical to be attentive to the professional and personality features of those individuals who are developing the strategy. Therefore, senior administrators at colleges and universities should intelligently select accreditation managers, committee or team leaders, and members of the self-evaluation teams. Team participants should be methodical in their work, detail oriented, analytical and evaluative in their perspective, and possess excellent report writing ability; however, they should not possess a biased, departmental, overly analytical, or indecisive approach, or have other agendas that may hinder the process. In addition, truthfulness and candor are also very important for accreditation members, and participants should be disposed to examine institutionwide issues and look for reasonable solutions that can be accomplished. (Most institutions in the United States are accredited by regional associations. See Appendix F for a sample selection of the expectations for a standard self-study report selected from the Self-Study Manual of the eastern Association of Schools and Colleges [ACCJC 2005].)

Attention should be paid to the workings of the self-study procedure so that the accreditation participants at the institution under review will remain focused on the specific intentions and study objectives with a sense of elucidation, consensus building, completeness, and priority (Young et al. 1983). Contributions to the self-study need to be continually inspected and refined, along with the college or university environment, agenda, and procedures. The accreditation standards should also be reviewed once the self-study has begun to justify and support legitimate criterion levels that are appropriate to the educational institution being studied and to review the material for committee members. It is important for the self-study process to use surveys or other information-gathering tools to aid in collecting facts as well as opinions, along with outcomes assessment procedures and their results, because these are now deemed necessary by most accrediting agencies. The workings of the self-study should also contain a plan to officially discuss the self-study results and plan for using this useful report in instigating important institutional improvements. The use of consultants from other colleges and universities who are peers is encouraged by some accrediting agencies prior to the start of the self-study. Of course, once the study has begun, visitation teams from outside the organization can provide valuable external perspectives individually and as a team to the institution for improvement.

Kells (1994) wrote a useful guide on self-study processes for postsecondary and similar service-oriented institutions and programs. The important essentials of the preparation and design of the self-study process include creating and using a prestudy planning group, establishing leadership, fostering internal motivation, establishing an agenda of local needs, designing the study process, planning for change and continuing study, and finally approving the design (Kells 1994). The prestudy planning group, which may become the steering committee for the actual self-study, has a very important function before the process begins. Planning is a part of management, and in accreditation management the plan for the self-study process is perhaps the most crucial step to achieving successful results. In addition to conducting the SWOT analysis (as discussed earlier), many important matters need to be discussed during the prestudy planning phase, including what type of self-study will be conducted (such as comprehensive or focused), project staffing decisions, timetable, funding sources, and other essential factors needed before the next step can begin. The prestudy phase can usually be accomplished in several meetings held over a month or term within the academic year, depending on the determination of the group leader.

Strong, positive leadership along with internal motivation and an appropriate design were found to correlate significantly with the perceived usefulness of self-study processes in a large study of institutional self-evaluations (Kells 1994; Kells and Kirkwood 1979). The analysis found that in those institutions where the leader, such as the chief executive officer, was enthusiastic about the self-study and made it a high priority, the respondents reported improvements and other useful results from the self-study. The prestudy planning group should assess the institutional leadership, and if needed, formulate a plan to gain this kind of support for the process.

Influential members of the prestudy committee may be able to convince the institutional leadership about the necessity of this public enthusiasm and organizational priority through personal meetings. The prestudy team could also ask the leadership to authorize a visit by a representative from the accrediting agency staff or a self-study consultant who could discuss the possibility of using the self-study to help accomplish the leader's preferred tasks or agenda items. Being that there are now several options for more than the traditional comprehensive self-study process by many accrediting agencies, leaders need to be informed of all the options at an early stage. Top-down organizational support and priority are critical for success in managing accreditation activities to achieve a variety of institutional goals. The leaders should show the educational community that the self-study process is not going to be just another worthless exercise that will achieve nothing important, but can benefit the entire community through recognition and rewards. This should be equally true for both regional and specialized accreditation programs. This can be done by words, actions, and leveraging the leader's strengths in appropriate ways to improve the motivation of the participants.

Both regional and specialized accreditation processes should be widely advocated internally within the educational institution to generate strong enthusiasm in faculty members, administrators, office staff, institution, trustees, alumni, and students. For example, one midsized comprehensive college in a metropolitan area of the United States recently attained specialized accreditation by AACSB International. For more than 20 years the college wanted this esteemed recognition and it was previously attempted without success, until a new leadership team came into place. The board of trustees, a new president, new chief academic officer, and new dean of the business school all made specialized accreditation a top priority for the institution in both their words and actions. The college president even went so far as to publicly proclaim at a collegewide meeting that the "train had left the station" (toward achieving accreditation), and if educational community members were not on board they would be left behind. In addition, the chief academic officer (provost/vice president for academic affairs) also stated publicly that the college would do "whatever it takes" to achieve this accreditation, and therefore the business dean was well supported in the accreditation process prior to and throughout the eventually victorious self-evaluation process. This example demonstrates that ongoing influence, officially and unofficially, by the institution's senior administrators of the accreditation planning group and the college itself is very important for producing and nurturing ongoing internal inspiration, especially in the early stages of seeking accreditation.

Proper planning and support from the institutional leaders is important for not only specialized accreditation pursuits, but also for the very important regional accreditation evaluations. A good example of effective self-study planning and preparation can be seen in the example of the New Mexico Institute of Mining and Technology (New Mexico Tech [NMT]), which is a public comprehensive institution that is one of three research universities in New Mexico (Yee and Zeman 2005). In the fall of 2004, NMT conducted the requisite comprehensive decennial evaluation for renewal of their regional accreditation by the Higher Learning Commission of the North Central Association of Colleges (NCA) and Schools. The NCA recently updated its Criteria for Accreditation, and NMT chose the ambitious option of using the new criteria, which were published only six months prior to the scheduled on-site peer-evaluation team visit. Prior to the self-study, NMT sent several institutional representatives to the regional association's annual meetings for several years to help prepare for the reaccreditation visit. The representatives attended as many sessions as possible, and in particular were attentive to sessions that addressed assessment of student learning. The new accreditation criteria and guidelines were explicit in their direction that institutional self-evaluations would now be more evaluative than merely descriptive. New Mexico Tech interpreted this to indicate that the institution should carefully examine what it currently did; whether it was effective; identify any corrections, improvements, or additions that were needed; and plan to document and quantify those findings. Each self-study chapter included a special section on specifically what needed to be addressed and how it would be done. The teams were appointed, research and drafting of documents were conducted, and a resource room was prepared for the accreditation peer-review team visit (Yee and Zeman 2005).

The institution was warned that a prospective timeline of only 18 months was not sufficient to properly conduct a self-study. Nevertheless, the institution found that although it was quite demanding, it was successful because the accreditation steering committee and subcommittee faculty members worked diligently on the project, even during the summer, and off-campus research was conducted. In addition, a faculty focus group meeting with a professional facilitator was conducted to discuss follow-up steps for the self-study, and that meeting produced plans and suggestions that were sent to the strategic planning committee. Although it was an aggressive timetable, the NMT self-study plan shows that a well-organized evaluation using proper preplanning with attendance at accreditation meetings, a visit by a liaison from the association to assist in the process, and the appointment of another external facilitator for follow-up steps to the accreditation self-evaluation can be worthy strategies for achieving accreditation approval and organizational renewal. The 18-month timeline for NMT is shown in table 2.

As with many large projects involving dozens or hundreds of individuals and numerous teams (committees), accreditation steering committees will often find an uneven level of effort or involvement by committee members and teams as a whole in the writing, editing, and participation in steering the accreditation process. One strategy suggested by NMT is to add faculty focus groups, student focus groups, and other activities to encourage solicitation of feedback during the writing process. Faculty engagement is important, and Part Four of this book explores some new options using electronic learning systems for enhancing team performance in accreditation self-study procedures. The results of internal and external self-studies, involving key faculty members and administrators at institutions, offer the opportunity for review of overall institutional effectiveness, and at NMT it laid the groundwork for the next strategic plan (Yee and Zeman 2005).

Another well-planned and complete self-study management procedure is illustrated in figure 4, and illustrates how the pre-accreditation self-study planning fits into a comprehensive representation of an effective self-study process within an accreditation management outline. This model is based on self-study timelines by Kells (1994) and Bartelt and Mishler (1996), with adaptations added by this author to integrate accreditation management objectives (Alstete 2004). The figure shows how the planning stage in accreditation management entails crafting a prestudy accreditation planning team roughly one and a half years before the scheduled

Table 2 New Mexico Tech Self-Study Timeline (Yee and Zeman, "How We Used the New Criteria," used with permission)

Months To Visit	Month-Year	Activity	Performed by	Months into Process
19	March 2003	Associate VPAA appointed as Self-Study Coordinator	President and VPAA	1
18	April 2003	NMT team attends the 2003 NCA Accreditation Association Annual Meeting	Steering Committee chair, member, and Self-Study Coordinator	2
18	April 30 2003	Tentative plan presented the NMT President and Cabinet	Steering Committee chair, member, and Self-Study Coordinator	2
17	May 16 2003	Steering Committee and tentative self-study plan announced at faculty council	President	3
16	June 2003	Steering Committee meets regularly and begins planning strategy	Steering Committee	4–6
16–14	July	Self-study Web page posted	Staff	4
16–14	June–August 2003	Steering Committee structure defined. Initial data and reports collected and made available. Self-studies from other schools collected Steering Committee		4–6
13	Sept. 2003	Criterion Teams formed and charged with tasks. Draft plan sent to the association	Steering Committee	7
13–12	Sept/Oct. 2003	Accreditation liaison invited to visit campus	President	7–8
13–10	Sept/Dec 2003	Assigned tasks being completed	Criterion teams	7–10
13–10	Sept/Dec 2003	Bi-monthly meetings and coordination of Criterion Team's work	Steering Committee	7–10
12	Sept. 2003	Tech Smart Quadrathalon	Steering Committee	8
10	Dec. 2003	Accreditation association liaison visits campus		10
9–7	Jan–Mar 2004	Compilation of Criterion Team reports	Steering Committee	11–13
7–3	Mar–July 2004	Draft of self-study, Drafts revised	Steering Committee	13–17
6	April 2004	TechSmart presentation at Association Annual Meeting	Self-Study Coordinator and Committee Chair	14
3	Mid-July 2004	Final draft of self-study	Steering Committee	13–17
2	August 1 2004	Self-study submitted to the association and prepared for regent's retreat	Steering Committee	17
0	Oct 4–6 2004	Consultant-Evaluator team visits campus		20

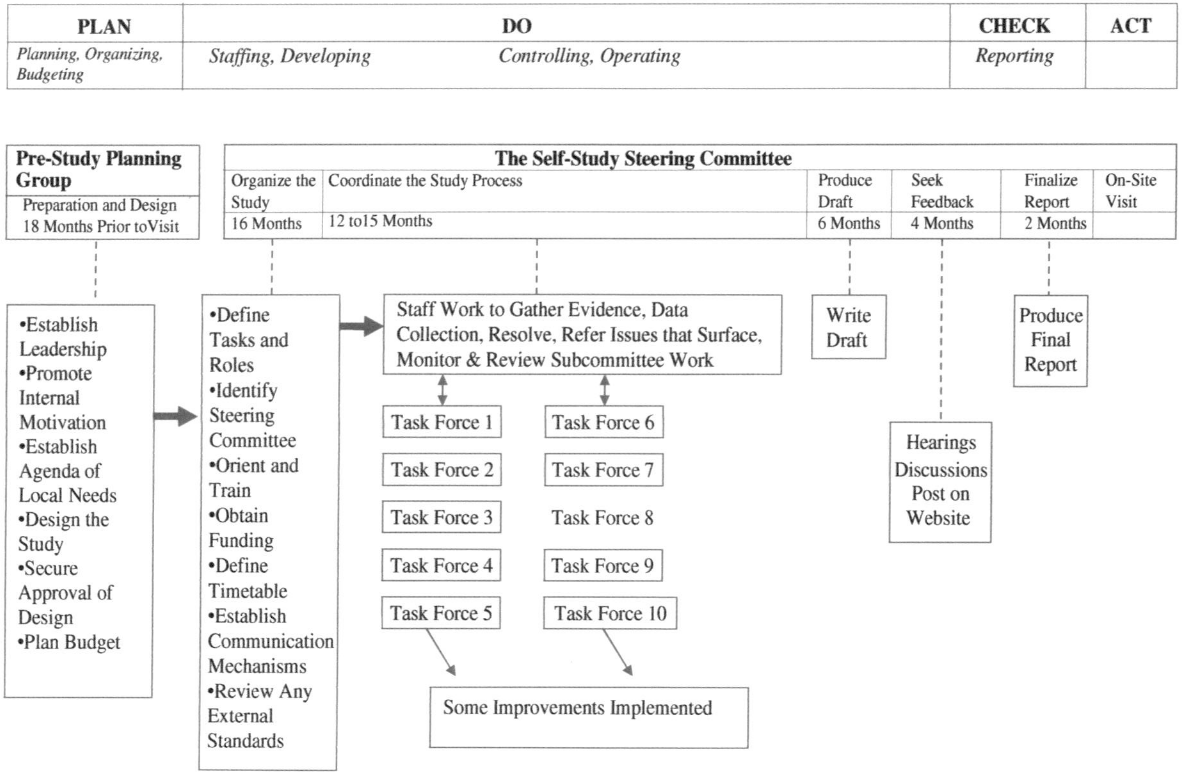

The Self-Evaluation Analysis • 65

PLAN	DO		CHECK	ACT
Planning, Organizing, Budgeting	Staffing, Developing	Controlling, Operating	Reporting	On-Site Visit

Pre-Study Planning Group	The Self-Study Steering Committee				
Preparation and Design 18 Months Prior to Visit	Organize the Study	Coordinate the Study Process	Produce Draft	Seek Feedback	Finalize Report
	16 Months	12 to 15 Months	6 Months	4 Months	2 Months

Pre-Study Planning Group:
- Establish Leadership
- Promote Internal Motivation
- Establish Agenda of Local Needs
- Design the Study
- Secure Approval of Design
- Plan Budget

Organize the Study:
- Define Tasks and Roles
- Identify Steering Committee
- Orient and Train
- Obtain Funding
- Define Timetable
- Establish Communication Mechanisms
- Review Any External Standards

Coordinate the Study Process: Staff Work to Gather Evidence, Data Collection, Resolve, Refer Issues that Surface, Monitor & Review Subcommittee Work

Task Force 1, Task Force 2, Task Force 3, Task Force 4, Task Force 5
Task Force 6, Task Force 7, Task Force 8, Task Force 9, Task Force 10

Some Improvements Implemented

Produce Draft: Write Draft

Seek Feedback: Hearings Discussions Post on Website

Finalize Report: Produce Final Report

Figure 4 Accreditation Management Framework for a Self-Study Planning Timeline

accreditation peer-review team arrives at the college or university under review. Leadership for the self-study process needs to be selected and appointed by the institution for this procedure and be advocated internally to secure enthusiasm regarding the accreditation goals. The academic community needs for the self-study management plan should be ascertained, followed by the design of the study. Proper institutional governance approvals should be obtained, and a budget needs to be designed. Effective accreditation management requires that all of these matters be included for getting a maximum return on investment (in both money and institutional effort in self-study accreditation process for the college or university. Performing a standard self-study often requires a timeframe of approximately one year to eighteen months, during which time the study plan is prepared and coordinated, and a written self-evaluation report document is fashioned.

The responsibilities and roles of the of the accreditation task force members need to be expressed in the planning of the self-study, overall steering committee members need to be selected and instructed on their duties, budgetary funding approved (normally founded on the accreditation budget planned in the first step), the accreditation timetable determined, member interaction procedures established, and the accreditation standards reexamined. The accreditation team members work to collect large amounts of information during this step in the self-study, organize the data, settle any problematic issues that arise, and scrutinize subcommittee operations to ensure validity.

When an initial draft of the self-study is produced, the checking stage begins. During this important step, all of the large work produced by the task force steering committee and subcommittees is verified for content correctness. Reaction is also sought from internal institutional constituents and external stakeholders in the community, often using printed and Internet publication of the draft self-study report, open institution meetings, and advice from the senior college leaders. The concluding step in a typical self-study procedure is to complete the self-evaluation report and take additional action on some of the report's conclusions before the accreditation team visit. These endeavors can be based on some of the potential or real troubles and problematic issues uncovered during the self-study process, with the specific follow-up actions consisting of drafting institutional strategies to solve the problems that were uncovered. While some improvements based on the issues discovered can be brought about immediately within the institution, most accreditation associations are looking to assist colleges and universities to systematize proper recognition of problems and be proactive, rather than reactive, in a continuous improvement fashion. The strategies and operational plans that are developed to solve the important and often difficult findings will normally result in attainment of the accreditation sought unless the institution does not address or plan to address the problems uncovered.

Finally, after submitting the self-study conclusions to the accrediting association, the last step in the self-evaluation procedure is to integrate the improvement plans and improved rotation of self-study and planning into the college's or university's internal organizational structure and culture as a foundation for long-term strategy planning and continuing institutional research. The education achieved by administration and faculty leaders within the educational community about the importance of self-evaluation and effective planning for such a process can be as important to the institution's progress as are the specific findings reported in the self-study document itself.

CHAPTER 9

The Peer-Review Visit

In the closing step of the self-study part of the process, the accreditation association uses volunteer peer-evaluators to review the self-study report and determine whether the evaluation team visit should be scheduled on-site at the college or university undergoing accreditation appraisal. This appointed team of qualified academic peers is dispatched to the institution. It evaluates how well the college or university meets the published standard of the accreditation and offers perspectives on the strengths and weaknesses of the institution. After the visit, the visitation team writes a written report to assist the institution in enhancing its academic programs and also outlines the criteria according to which the accreditation association decides to approve of, continue or reaffirm, or deny accreditation authorization (SACS 2002).

It is in the course of a standard two- or three-day visit that the visiting accreditation team members inspect institutional data and perform in-person interviews in order to assess the value and accuracy of the self-study report that was submitted and discover whether the college or university has fulfilled the accreditation standards of the specialized or regional agency the team is representing. The visitation team usually tenders written suggestions to the institution, builds a team consensus on their results, and writes the draft of the official report. Lastly, the visitation team frequently offers a verbal synopsis to the vice president of academic affairs, dean, or chief executive officer in an exit statement with other invited college or university officers on the final day of the stay. It should be known that the departure of the visiting team after the visit does not signify the conclusion of the accreditation evaluation. Normally, the visiting committee's evaluation report and the response of the institution to the committee's findings are reviewed by the accreditation agency, and then a final accreditation decision is completed. The participants on a review team serve somewhat of a dual role as respected equals offering consultation and as evaluators. The goals include producing a written team report that will be practical to the college or university and to the accreditation agency that must compose an official decision on accreditation.

An on-site visitation committee typically contains 5 to 15 people, and the specific total is based on the characteristics of the college or university and its academic offerings, as well as on the accreditation agency's standard operating procedures. Generally, every primary academic instructional area that is under review by the accreditation association must be inspected. The on-site team members are chosen from similarly accredited colleges and universities that are also accredited by the association, suitable organizations, and occasionally different accreditation agencies or academic subjects. As mentioned, institutions seeking accreditation often endeavor to have some of the on-site reviewers come from colleges and universities similar to the one that is being reviewed. Most of the visitation team peer-reviewers are selected from colleges and universities that are usually some distance away, such as in a different state or region of the United States from where the institution being reviewed is located. The accreditation agency typically has the final say on visitation team member assignments and structure. Knowledgeable reviewers from other areas or from other established accreditation associations are then officially assigned to the on-site visitation teams. For impartiality, evaluation committee participants are usually chosen who have no previous or current relationship with the college or university being reviewed, and are also not believed to be local rivals.

The on-site evaluation team chairperson is appointed by the accreditation agency well in advance (usually more than one year) of the scheduled campus visit, and this leader is frequently a current or former institutional member of the accreditation agency. When choosing the on-site evaluation team, effort is usually made by the accreditation agency to circumvent possible conflicts of interest—for instance, when an on-site team reviewer had a previous position as an employee of the college or university being reviewed or may have other individual prejudice that is unfairly critical of the institution. The approved on-site evaluation team membership list is sent to the institution well before the scheduled visit, and the chief academic/executive officer or school/department leader is instructed to inform the accreditation association headquarters if there are any concerns about the makeup of the accreditation assessment team that has been assigned.

The accreditation association organizes the on-site team visit schedule often largely in coordination with the college or university president's office because of the leader's very busy calendar, and this is frequently done up to two or more years prior to the visit (NASC 2002). The accreditation agency makes significant efforts to schedule on-site visitation dates that are most suited to the institution's needs. Nevertheless, as is common in such matters, concessions on the requested dates are occasionally necessary to accommodate various conflicts, such as forthcoming accreditation association meetings. Most accreditation associations expect the on-site team visit to be from two to four days, in total. This on-site evaluation is often planned for weekdays or a mixture of weekdays and weekends, based on the current requirements of the accreditation agency and the college or university being evaluated (AAM 1989). A representative daily schedule for an on-site campus accreditation team visit for U.S. regional accrediting association is shown below:

Higher Education Accreditation Commission
Onsite Evaluation Team Visit to Intermediate-Size University
November 11 to 15

SCHEDULE
Tuesday, November 12
5:00 pm to 9:00 pm

Welcome, Cocktail Party and Dinner
Evaluation Committee, Chancellor/President's Cabinet,
University Accreditation Steering Committee Members
Place: Prestigious Dining Club

Wednesday, November 13
9:00 am to 10:15 am
Evaluation Team Orientation
Place: Board Room, Chancellor's Hall

10:30 am to 11:15 am

President's Cabinet
Chancellor's Hall
Dean of Libraries
Director of Mission Integration
Place: Board Room, Chancellor's Hall

11:30 am to 12:45 pm

Interviews
Place: Conference Rooms

1:00 pm to 2:00 pm

Steering Committee (lunch)
Place: Board Room, Chancellor's Hall

2:15 pm to 3:15 pm

Faculty Senate
Place: Faculty Reception Room

3:30 pm to 4:15 pm

Open Walk-In Session
Place: Board Room, Chancellor's Hall

Thursday, November 14
8:00 am to 9:15 am
Members of Legal Board of Trustees
(breakfast)
Place: Board Room, Chancellor's Hall

9:30 am to 10:45 am
Interviews
Place: Conference Rooms

11:00 am to 12 noon
Students (brunch)
Place: Student Lounge

12:15 pm to 1:00 pm
Open Walk-In Session
Place: Board Room, Chancellor's Hall

1:15 pm to 3:00 pm
Interviews
Place: Conference Rooms

3:00 pm to 3:45 pm
Department Chairs, Directors
Place: Board Room, Chancellor's Hall

Friday, November 15
11:00 am to 12:15 pm
Chair's Oral Report to the University
Community
(Students, Faculty, Staff, Administrators
Steering Committee, President's Cabinet)
Place: Auditorium

4:15 pm to 5:30 pm

Interviews
Place: Conference Rooms

12:30 pm to 1:00 pm
Chair's De-Briefing Meeting with President
Place: President's Office

12:30 pm to 1:30 pm
Optional Lunch for Team with
President's Cabinet
Place: Board Room, Chancellor's Hall

A primary objective of the team visit for the college or university being evaluated should be to not only put forward the institution in positive terms, but also to ease communication with a diverse intrainstitutional community as much as possible for useful information gathering by the visitation team. If the self-study is a concentrated review that focuses on a specific area that the college or university has selected, then the composition of committee members as well as the daily visitation schedule may be different from the sample representative schedule shown here. Nonetheless, the plans for many accreditation team visits normally include data gathering through personal interviews with community members and comments given at the finale of the on-site visit.

CHAPTER 10

Publication of the Evaluation

Often, the visiting committee chairperson is responsible for editing the individual reports and publishing a confidential committee report for the accrediting association during the on-site visitation or shortly thereafter. Since the committee chair usually has final authority for the content of the report, it is important for the educational institution to pay extra attention to the individual selected as the visiting committee chairperson. In addition to strong writing skills, the visiting team chair often has strong leadership and team management skills, and the ability to use information technology to produce an effective visitation report in a timely and accurate manner.

The actual evaluation report is usually a planned, reasoned document written for and addressed directly to the educational institution and its constituency, clearly presenting the major findings and suggestions of the evaluation team. It is a critique by competent external colleagues for an institution's own use. Such reports have an important function in the accreditation process in that they assist the agency in its decision-making processes, but the primary users of the reports are the institutions and their constituencies. Normally, the accrediting agency that writes the report considers the evaluation team's report to be the institution's property.

The purpose of the accreditation evaluation report is to validate the institution's self-study and make recommendations to the institution on the ways in which it can improve its effectiveness. As previously mentioned, accreditation team members are drawn, in most instances, from the staffs of other accredited institutions, and normally guided by an experienced chairperson. The members who write the report should be oriented through workshops and certain published accreditation agency documents, but the agency usually does not attempt to give the team specific formulas or blueprints. Accreditation team members rely upon their own knowledge and observation of academic excellence, applying this knowledge and experience within the context of the agency's eligibility requirements and accreditation standards.

Therefore, accreditation agency reports should not be expected to uniformly express the same views, or be in all respects consistent with one another. Even when a report has been approved by the agency, it is still an expression of the views of a particular group of educators on questions that are often too complex to have single answers. A different team might have seen things somewhat differently. The accreditation agency typically accepts such possibilities as the natural consequence of the developmental and often experimental nature of higher education, of its commitment to diversity rather than conformity, and of its conviction that it would rather trust the judgment and influence of professional colleagues than of a small, employed staff of full-time "experts."

An example from the guidelines of a common regional accrediting agency recommends that educational institutions should be governed by two principles in using an evaluation report (MSACHE 1995):

(1) The evaluation report should be studied open-mindedly and seriously by appropriate constituencies, because it is the thoughtful product of sensitive, disinterested professionals;
(2) The educational institution must respond fully to team recommendations relating to compliance with agency standards; however, the institution reserves the right to accept, modify, or reject team suggestions or advice.

To the extent that an accreditation agency's evaluation team report is advisory, it is more the basis for further thought than a statement of a final conclusion. This concept is in keeping with a collegial and scholarly mind-set common in the postsecondary community, and leaves room for negotiation, adjustment, response, dialogue, and growth. Accreditation agencies may also have an official policy on collegiality and public communication, where the educational institution is required to make the report readily available or distribute it as widely as possible on campus, since the report is addressed to an institution's entire constituency—administration, trustees, faculty, students, and staff. When distributing the report, however, the institution should indicate that the report does not constitute a summary of the entire accreditation agency evaluation process; it is only the report of the team that visited the institution. Since the agency's review processes sometimes result in an accrediting action other than the one recommended by the team, misunderstandings may occur if it has not been made clear that the report is only one piece of a much larger whole that includes the institutional self-study, the site visit, the accrediting agency committee review, and deliberations of the full accreditation agency decision-making body.

Aside from its internal constituencies, the institution is normally free to distribute copies of the report to others at its discretion. Should an institution use the report in such manner as to create a misleading impression—for example, by using selected excerpts—accrediting agencies typically reserve the right to release the full report and to make appropriate statements to the public. Excerpts, when used, should be verbatim or reasonable paraphrases and must accurately reflect the entire report in its balance of strengths and team concerns. As part of the accrediting process, confidential copies of the evaluation team report may be distributed to the members of the evaluation team and heads of multiunit and regional systems, and in joint or cooperative evaluations with the chief executive officers of the other accrediting agencies involved. Accrediting agencies for most academic associations do not share evaluation team reports with government, public, or private agencies or individuals unless explicitly permitted in writing to do so by the institution.

Although there may be some disagreement with parts of it, the report is normally aimed at improving the institution. In order to achieve this goal, the trustees, administrators, and faculty members must study and seriously consider it. The visiting committee report is not a mandate to the institution, but is more of an important advisory tool. It is understood and accepted that the institution has the right and obligation to plan its own course of action, and that action may or may not be in full agreement with the suggestions and recommendations of the evaluation committee. Most accrediting agencies expect the institution to use the report objectively and in good faith. The report can also sometimes be used for internal support and external publicity purposes, but in preparing public announcements the educational institution should check with the accrediting agency on this, and may need to avoid quoting directly from the report or reporting only certain favorable or unfavorable findings.

CHAPTER 11

Following the Accreditation On-Site Evaluation

After the conclusion of on-site evaluation and publication of the official institutional evaluation, accreditation agencies could expect or allow that the college or university leaders write a detailed document to address the concerns raised in the accreditation evaluation report (MSACHE 2001). The accreditation associations might additionally necessitate a second on-site visit to an institution if it is warranted, and this could be done following an analysis of the official institutional follow-up report or at the behest of accreditation members because of the institution's shifting situation or in the external surroundings. Accreditation associations often call for a college or university to prepare a follow-up document when the association requires additional data on a particular topic that was not sufficiently addressed by the institution's cyclic self-evaluation report, a first accreditation self-study report, or in answer to a suggestion that was proposed by the on-site visitation committee for the association. The association's accreditation director or representative may seek to help the college or university by concentrating initial awareness on a singular issue of interest or one that requires additional information.

As with the self-evaluation report, the research for a follow-up accreditation report can present occasions for productive conversations on a college or university campus that include many stakeholders of the institution's community and give rise to many perspectives on a specific subject. The follow-up report can also function as a useful planning mechanism for the institution because it has external validation and significant institutional backing. The layout for this type of document normally has no predetermined format, except that it should include a purposeful title together with the college or university's name and location; the date on which it was written or finalized; and an important citation from the accreditation association's letter that asked for the follow-up report, extracted appropriately so that it distinguishes the focus of the document.

The specified issues and follow-up activities will influence the scope of the follow-up report because they were in the agency's letter that led to the accreditation action. If there are questions that the college or university has about the accreditation agency's concerns, someone from the institution who understands the accreditation process can communicate with an association staff member, who functions as liaison with the educational institution. The content of the follow-up document should create the framework for the information being offered, and since at least one or more years normally intervene between the accreditation association's request for this follow-up report and the specific date on which the document is submitted, the document content should go through adequate institutional conditions to make the magnitude of recent developments apparent to readers who may not have been involved with the institution's accreditation issues. The follow-up report document can be as concise as a one- or two-page memo to the accreditation association. Nevertheless, the report should be as clear and specific as the content of the institutional information allows, thereby enabling readers to measure the condition of the institution to arrive at an informed conclusion.

Although being concise and directing the report to the issues is important, it may be appropriate to also supply relevant supporting information if the data fortifies and elucidates the main points of the document. Any appendices or supplementary items should be limited to only those materials that are necessary to the accreditation agency's reading of the report as it pertains to the requirements of the follow-up. This additional documentation should not be used as more unnecessary selling of the institution, but should instead be focused on interpreting and presenting information so that its application to the matters being addressed is clear.

Once the college or university submits the follow-up report, the accreditation association may direct a review committee of commissioners to examine the follow-up activities for institutions that are being considered for accreditation or candidate institutions to review each report document. Additionally, follow-up reports that contain budgetary information may be analyzed and deciphered by an accounting expert. Once the document has been read and understood at the agency, the next step is that the accreditation association will determine which one of the possible actions to implement that are delineated in the choice of actions available. Additionally, follow-up accreditation team visits are sometimes scheduled subsequent to a definite action of the accreditation association or at the request of a college or university under review. The accreditation association moreover may compel a visit after examining a follow-up document or additional information presented by accreditation personnel. The topics or problems examined in a special accreditation follow-up on-site visit normally are restricted to precise issues. Nevertheless, as regional accreditation in the United States pertains to a whole educational institution, and professional or specialized agencies apply to a particular academic area, the accreditation follow-up review entails an assessment of the individual issue within the framework of the educational institution or specialized program. The composition and size of the follow-up visitation team may be smaller than the first on-site committee, and some agencies add in an association staff official observer and/or a delegate from the appropriate state government education office. It should be noted that these follow-up on-site accreditation visits are not the typical reaction by accrediting associations for normally periodic or initial regional accreditation evaluations. However, the agencies customarily set a range of accreditation verdict outcomes that are possible for colleges and universities seeking accreditation.

These alternative judgments made by accreditation associations often include various stages of potential results. After educational institutions that were already accredited have gone through an evaluation by a normal regional U.S. accreditation association, the organization may choose from a list of official decisions stated in their policies. An agency normally provides written notification of this action to the educational institution within a specified time period of the date the action was taken (NWCCU 2005). Although the decision-types vary somewhat among the regional and specialized agencies, they generally follow a similar pattern, ranging from approval to rejection. One regional accreditation agency's list of actions with regard to college and universities includes:

1. Grant Candidacy or Initial Accreditation
2. Continue Candidacy or Reaffirm Accreditation
3. Request a Progress Report and/or a Focused Interim Report and Visit
4. Defer action on Candidacy or Accreditation
5. Issue or Continue Warning
6. Impose or Continue Probation
7. Issue or Continue a Show-Cause order with Candidacy or Accreditation to terminate unless the institution has demonstrated, to the satisfaction of the Commission that it has satisfied the Commission's concerns or responded to its directives prior to a specified date
8. Deny Candidacy or Accreditation
9. Terminate Candidacy or Accreditation. (NWCCU 2005) (used with permission)

If follow-up reports from an educational institution or special visits are required as the result of action taken by the accrediting agency (or in the case of a candidate institution following a status review visit), such institutional reports and visitors' reports are reviewed directly by the agency. The accrediting agency then normally takes an action parallel to those listed above for renewal of accreditation or initial accreditation, except that if accreditation was reaffirmed at the time the follow-up activity was required, reaffirmation is not repeated.

Supervision by accreditation associations generally entails institutional periodic review reports that must be submitted every few years, and some professional or specialized agencies are now demanding annual information reports to be submitted electronically by accredited colleges and universities for ongoing assessment and evaluation. For example, AACSB International has a Knowledge Services office that coordinates information and data about the individuality and procedures of business colleges and their activities to strengthen specialized business accreditation, support institutional development, improve accreditation member services, increase awareness by the general public, and assist with informed accreditation evaluations. However, some accredited college and university officials find these recurring annual information survey reports to be very time-consuming and probably not immediately valuable for internal development in the short-term operational planning or even for their long-term strategic decision-making exercise. The annual online data-collection procedure required by AACSB International is intended to gather comprehensive data and information that classifies and illustrates members' academic business units and their relationships within the larger institution. Business colleges and schools completing the Internet-based annual questionnaire can obtain a sequence of wide-ranging statistical reports that analyze the broad data set collected from AACSB International members, gain the ability to design and buy specially requested reports, and obtain information outlines of their business program or school on the AACSB International Internet site for additional recognition or publicity. Other specialized accreditation associations may also start requiring annual data-gathering soon, and some have proposed more radical developments away from the traditional prearranged periodic evaluations of seven- or ten-year cycles, to ongoing annual evaluation with no on-site campus committee visits unless the need is decided. This evolution may be partly due to the growing influence of internal information systems at organization, the growing power of the Internet, disparagement of traditional accreditation efficacy, protests about the rising costs of educational institutions pursing accreditation, and other issues. However, colleges and universities need not be overly concerned about potential changes at this time, because the currently required accreditation procedures must be properly planned and addressed by educational organizations now, apart from what future developments may occur. Consequently, academic administrators and faculty members at colleges and universities today ought to consider many of the alternatives for effectively managing the accreditation processes for superior performance so that the institution's strategic goals and objectives can be attained.

PART THREE

Managing Accreditation at Colleges and Universities

CHAPTER 12

Accreditation Management

The preceding chapters have illustrated how the environment of postsecondary accreditation is swiftly evolving and is quite unlike that of the early era, when small, independent colleges needed little public recognition and seldom focused on internal renewal. Today, institutional leaders, senior and midlevel administrators, faculty members, and even staff are often required to collaborate on frequent accreditation projects with quasi-governmental associations that bestow official identification as well as great occasions for internal self-renewal. This will happen only if the institutions and their senior administrative leaders choose to support and direct these opportunities and lead the quest for early reaffirmation of regional and specialized accreditations as more than just a duty that must be fulfilled. These days, most postsecondary institutions are under strong pressure from internal and external forces to use assets as economically as possible because competition and other issues that are affected by national and international forces are having a significantly larger role in configuring higher education service conveyance. Consequently those employed in the field of postsecondary education must proactively manage these developing responsibilities and challenges intelligently. The management of accreditation procedures is a crucial duty that should be designed and performed to supersede the demands of these growing market forces and global challenges. The reader may wonder what precisely is intended by accreditation management and how this concept corresponds with the structure and goals of postsecondary administration today.

The relevant scholarly writings and research on administration and management offer several viewpoints on how these two concepts are dissimilar and how management is normally consigned to a secondary and particular subsection of overall organizational administration processes (Lynn and Stein 2003). A classic management scholar, Henry Fayol (1930), stated that it is essential to not mistake administration for management (Fayol 1930). Fayol postulated that to manage properly is to run an organization toward the greatest potential use of all the assets at its availability and to guarantee the efficient functioning of the organization's critical purposes. While true administration fulfills only one of these roles, it is defined as the universal process of efficiently organizing people and directing their activities toward common goals and objectives (Simon et al. 1991). For colleges and universities, administration can consequently be understood as an elevated level of supervision in the formation, growth, and execution of organizational goals in pursuing and renewing accreditation by regional and/or specialized accreditation associations. Management of accreditation processes at postsecondary institutions, soon to be expressed and developed here, will be revealed as the valuable methods by which colleges and universities utilize all of the assets and capabilities they have on hand to attain internal organizational revitalization and external recognition from accreditation associations and society.

Just as this book has documented the rapid developments in higher education accreditation and postsecondary education so far, the thinking in the field of management of organizations has also developed quickly over the recent years. However, the roots of administration and management theory are quite older. In the mid-nineteenth century, the phenomenon of management as we know it was unknown (Drucker 2001). Corporations, public and nonprofit organizations, and especially colleges were very small by today's standards. In less than a century and a half, management progressed and changed the economic, societal, and governmental panorama of the many modern and leading countries in the world. Classic management theorists such as Fredrick Taylor, Max Weber, Alfred P. Sloan, Elton Mayo, W. Edwards Deming, and others created several perspectives to describe and comprehend management theory from academic, illustrative, comparative, and action-based viewpoints. The models acknowledged, clarified, and disseminated by these management theorists helped to grow and improve many types of institutions and organizations in addition to business corporations, such as public organizations including government agencies, nonprofit institutions, and higher education institutions.

Even though management is known to be very essential for success, it is still one of the most fragile yet ubiquitous activities in societies today (Hardin et al. 2005). Management roles and responsibilities are carried out in most of the everyday lives of individuals, as well as in government offices, religious congregations, and, especially, the business decisions of corporations. Yet historically, management has not been the focus of organizational development and evolutionary growth. This is because the focal point has been on product or service production, supervision of the organization's employees, or some mixture of both. Nonetheless, effective management strategies have been and will remain to be a significant benefit for institutional managers and leaders. By considering the style and achievements of great leaders from the past, a reader can comprehend that most or all achieved great expertise in their utilization of management techniques, such as organizational planning and directing of people. The administration and especially management of institutional performance greatly influence the achievement and longevity of all types of organizations. Comprehending the historical and fundamental principles of organizational behavior is critical to accreditation management and success in higher education administration. The level to which the senior administrators (and governing boards in higher education) can understand and implement unassailable management performance mirrors what the organization will develop into (Geroge Jr. 1972).

This author believes that management abilities, organizational behavior, and leadership skills are vital to institutional survival and that an exploration into the background and history of management concepts will assist a deeper comprehension of the issues surrounding effective accreditation planning and supervision. College and university leaders should often review the events and conditions of historical precedent to fully comprehend where their organization seeks to go. The earliest human civilizations must have utilized some form of management and directorial systems to ensure their continued existence (Hardin et al. 2005). The classic works of Socrates argued about the talents of a good leader and the meaning of those skills to the social order. Nevertheless, the initial readings and exploration of management and organization started with classical organization theory in the eighteenth century.

The Roots of Management and Organization Concepts in Relation to Accreditation

Classical or established organization theory centers principally on the configuration of institutions and organizations. Leaders and managers who subscribe to this dogma often use rigid approaches to arrange the organization and argue that the finest decision is one that entails doing what is best for the organization. The idea here is that there is only one best way, and this principle triggers many decisions by managers and leaders, because they are understood to be realistic and decided for the best intentions of the organization. To add to the historical background of management, Shafritz and Ott (2001) state that the management practices of Moses actually provide evidence that organizational behavior and management techniques have been used for millennia. As written in the scriptures, Moses decided to use management by exception, (where managers only intervene where performance is a problem and there is much delegation) which is a practice that was not identified as such until many centuries later. A sequence of events in the eighteenth century then greatly affected the evolution of organization behavior theories. These events included the movement of the human population to larger cities, an increased specialization of human activities for employment beyond farming, the propagation of the printing press (which helped increase education levels), and the initiation of the Industrial Revolution.

While the business and management literature does not clearly identify a starting date for the beginning of organization theory as it is known today, most scholars believe that it began in the mid-1700s. In 1776 Adam Smith published *The Wealth of Nations* and was one of the earliest important scholars to examine the field of organization theory. This date in history is considered to be the operationally defined starting point of organizational theory as an applied science and academic discipline (Shafritz and Ott 2001). In his book, Smith writes about a company that manufactures pins, and shows tremendous understanding of basic management principles. Smith wrote about important and far-reaching concepts for his time, such as support for the division of labor in organizations, which he saw as crucial to the financial success of organizations. His thoughts on this and other concepts earned him a place in American history and the title "the Father of Economics," and his pioneering thoughts shaped the foundation of the liberal leadership approach.

Adam Smith supplied three causes as to why the division of labor was such an important procedure. To begin with, it permitted a setup where there is an increased competence in the abilities of individual workers; second, it reduced the time required for manufacturing because individual workers did not change functions; and third, it amplified the use of machines that were available to help the organization (Shafritz and Ott 2001).

By the nineteenth century, there was a larger quantity of theories and perspectives that were written with regard to the management of an organization (Hardin et al. 2005). In fact, the word "manager" was initially coined during this era, as well as the expression "job description." This was the term that was advanced as a means to create and record the specific abilities that were required by managers. One early and visionary manager of his time was Daniel McCallum, who performed an important function in documenting the duties of the manager. He obtained frontline experience for his managerial duties as the superintendent for the New York and Erie Railroad. McCallum discovered solutions to the challenges facing the railroad industry by utilizing Henry Poor's basic principles (who created information communication—a data-switching bank for the U.S. railway system—and revived the team spirit of the managers and workers) for managing large organizations. McCallum used these ideas to develop specific codes and rules that all employees would adhere to, and he believed that by establishing a division of labor and enabling certain people to have power, along with providing written documentation of organizational topics (similar to accreditation or other periodic reports), a large company could flourish (Geroge Jr. 1972).

The challenges of expansion experienced by the railroad began to be felt in other organizations, and just prior to the great expansion of higher education in the United States. There were similar clashes in the organizational cultures and working environments in both industry and postsecondary education. The general changes in the workplace were greatly affected by the movement of people to the cities, and leaders of the developing companies realized that the supervision and management techniques that they had used for a long time were not suitable any longer. These leaders and managers recognized that help was strongly needed to properly address the challenges of these growing organizations. This development of business size and organization preceded and prepared for the scientific movement in management (Geroge Jr. 1972). This movement was pioneered by Frederick Taylor. During this era, Henry Metcalfe also started exploring the management chain of command at the Franklin Arsenal, which required help in distributing and controlling assets. Metcalfe recognized that conventional management techniques were not working effectively and created a comprehensive control plan based on his perception that power should come from a central position. In planning control from a centralized position, written reports of organizational expenses enabled managers to review records for improvement much more effectively (Shafritz and Ott 2001).

The real-world familiarity of management pioneers like Towne and Metcalfe initiated a new age and tremendous growth in the field of organizational theory. Another classic theorist, Frederick Taylor, created his now-famous scientific management notion, which states that managers should be more than authoritarian rulers. Effective managers should grow and implement far-reaching ideas of management that include the doctrine of planning, organizing and controlling. These will become fundamental to the proposed accreditation management strategy for colleges and universities. Taylor's interpretation of factory activities, its employees, and working environment, in addition to his basic time and motion studies, assisted the progression of his scientific management principles. Effectiveness and organizational control would increase revenues and profit if this standardization were applied, according to Taylor (Hersey and Blanchard 1988). Taylor's innovative thoughts have had an important influence on organizations and the culture they participate in.

It was also in the same era that Taylor advocated scientific management that Henri Fayol created the first comprehensive theory of management. Although widely ignored by Americans, the French engineer Fayol promulgated a theory that included a strategy of management pertinent to all types of organizations. Where Taylor realized that effective management requires learning at all levels, Fayol developed principles of management consisting of planning, organizing, commanding, coordinating, and controlling. Later in the 1920s, Max Weber examined bureaucratic establishments and, in contrast to Taylor, considered the social influence of the bureaucracy on individual employees. Weber created a definition for the term *bureaucracy* and his writings supplied an understanding into the social and psychological effects of a bureaucracy on people. The term is often applied in higher education in reference to the administration or governance structures; it sometimes identifies a particular structural chain of command—the supreme bureaucracy—that managers and employees must either navigate or find an alternative route around. Weber's beliefs about human relationships in management are the foundation for many management theorists today. His explanation of the roles of individuals in organizations can be successfully applied today because this framework of managerial hierarchy stays in place (Hersey and Blanchard 1988; Shafritz and Ott 2001).

Institutions of higher education and other organizations usually need a good number of employees to thrive, and the separation of duties continues to be an essential feature in organizational philosophy. Later in the 1930s, Luther Gulick created the idea of using a set of administrative principles in managing organizations, and these ideas are known by the mnemonic term POSDCORB. Managers who believe in Gulick's work have a list of all the required abilities that a leader needs to manage an effective organization such as a business or educational institution. POSDCORB refers to the following terms: planning, organizing, staffing, directing, coordinating, reporting, and budgeting. These beliefs continue to be used by effective organizations (Shafritz and Ott 2001) and will be linked to the management of accreditation.

The organization and governmental leaders of the past centuries bought into classical organization theory, which was somewhat prescriptive because organizational decisions were based on interventions designed to advance the whole institution. The needs of the worker were not considered relevant or important when compared with the needs of the company. Unfortunately, this approach to management continues in modified forms today in many organizations, and it was not until the evolution of resource concepts began to center around employees that organizations even considered the needs of the workers. Critics of classical organization theory have stated that it tends to perceive the workers as controllable machines that are to be supervised. In response to these concerns, several management theorists broke from the classical tradition and developed what is now called neoclassical organization theory. This concept builds on classical theory, but is wider and not completely separate from the classical approach. Instead, the neoclassical theory adds a more human touch to the management science of organizational development concepts. Managers are positioned as coaches instead of mere authoritative directors, and this concept's major points are centered on employee or worker inspiration, justification, and creativity. Neoclassical theorists include Chester Bernard, Robert K. Merton, and Herbert Simon.

In keeping with the humanistic concepts, Robert Merton examined the nature of organizational bureaucracy, and especially Max Weber's "ideal-type" of system. Because bureaucracy is frequently seen as mechanistic and dehumanizing in an organization, Merton determined that Weber's bureaucracy applied a force on managers to become synchronized by policies and procedures, thereby making them methodical or organized to a fault. This methodical performance causes the administrative bureaucracies to be perceived as rigid, even though authentic bureaucracy is effective for accurate record keeping. This obsession with documentation does not eliminate overindulgence in the management of details, which actually can manifest in bureaucratic organizations or any other system. Administrative bureaucracy (whether referring to accreditation management or other examples) is then characterized not by the presence of error, but by the unfortunately negatively perceived ways that a system seeks to increase efficiency and reduce errors (Pugh 1990).

In rejecting the claims of Fayol, Gulick, Taylor, and other classical theorists, Simon criticized the way that the scientific method of management addressed the basic principles of classical theorists, in particular, its views that organizational efficiency is increased through specialization, that centrality is the most effective means for following directives, and that the authority of organizational control should be restricted to selected leaders. Simon postulated that these concepts could be shown to be contradictory and inefficient, fundamentally clashing with classical theory (Shafritz and Ott 2001). Much later in his career, Simon stated in 1991 that four principles contributed to the successful or unsuccessful functioning of organizations, namely, authority, rewards, identification, and coordination.

In regard to the human relational elements of effective management and creating a positive attitude in employees, managers must understand the basics of motivation. Research by Abraham Maslow explains the fundamental needs that motivate an employee to function. Maslow identified specific needs that take priority over others, and he labeled this precedence as the hierarchy of needs. Naturally, physiological needs are the most basic, followed by safety, belonging, esteem, and self-actualization. The more fundamental needs in Maslow's hierarchy must be addressed initially before the higher needs can be fulfilled (Boeree 1998).

Aside from Maslow, other motivational management theories have surfaced, such as Douglas McGregor's Theory X and Theory Y, which he wrote about in his book *The Human Side of Enterprise* (McGregor 1960). According to this theory, there are two ways in which employee motivation can be viewed, and both of these theories begin with the principle that the function of management is to gather the parts for production (together with people) for the financial advantage of the company. However, there are also great differences between the theories, which McGregor interestingly named as such to avoid descriptive labels. Theory X states that employees and people in general are lethargic and will evade work because it is in their nature. Contrary to this outlook, Theory Y proposes that people are productive and actually take pleasure in work. These management premises have also been identified by other management writers as autocratic or participative, and more basically as hard or soft management (Chapman 2005).

Management and Administration as Applied to Higher Education Accreditation

Despite its production-oriented and institution-based foundations, readers may be interested to know how far management organization theory has developed in the past century. Beginning with Taylor and Fayol in the early 1900s, followed by McGregor's Theory X and Theory Y just half a century later, organization concepts expounded various divergent ideas, which included changes in the view of the institutions as dominant in production of goods, to human employees being the most significant factor. While the management theorists largely focused on organizations in the private sector, the area of public management evolved from political objectives to seeking the addition of perceived value by the public. Generally, public management means performing certain tasks related to policy implementation in publicly supported programs (Jones et al. 2001). There is a large area of overlap between business and public management, particularly with regard to economics, organization theory, human resources, finance, accounting, and information resources. However, public management is not exactly higher education management and certainly not the same as private business management, even though a large portion of the postsecondary funding comes from publicly funded sources. Higher education management and administration are unique for many reasons, including the variety of stakeholders to be satisfied, organizational structures, mission, personnel issues, and other factors. As mentioned earlier, higher education has also been undergoing many of the same internal and external pressures to improve performance that private companies and government agencies are undergoing.

Corporate organizations, nonprofit government agencies, and higher education institutions have endeavored to tackle these challenges for enhancement by utilizing a number of popular systems that were originated in the management field. These other management methods include schemes such as planned programming budgeting systems, management by objectives (MBO), zero-base budgeting, strategic planning, benchmarking, total quality management, and business process reengineering (Birnbaum 2000). In adapting these management ideas, some have viewed these approaches as improvements if they are used and preserved; others, mainly detractors, have viewed them as a trend or whim that is just part of a predictable life cycle of conception, running, and eventual desertion. Whether existing and new management concepts adapted from business do become part of the education establishment or are abandoned, the basic tenets of business and public management have been and will remain a part of educational organizations for practical necessity and to meet the ever-increasing challenges that most institutions are now facing.

Aside from implementing innovative management techniques, many companies and other institutions have been undergoing internal changes resulting in leaner, less-hierarchical institutional structures that are focused around a set of broad, value-creating processes and particular competencies (Jones et al. 2001). The traditional rigid containers or departments of management within organizations have recently been exchanged for control systems that spread out and decentralize organizational decision-making so that the results go down in the company to where it is required, such as the point of sale, delivery, or manufacturing (Simons 1995). However, in colleges and universities where decentralized, layered management structures coexist with a sizeable, flattened element of faculty governance structures, the transfer of administrative and managerial decision authority to yet subordinate constituents may not be achievable without intense direction, personnel education, and restructuring—particularly if this entails management of higher education accreditation, which crosses over academic departments and schools contained within bigger institutions. Consequently, the issue as to how academic accreditation can be managed successfully within a structure that has significant organizational restriction and still expand the true capability that systematic management strategies can provide still remains to be discussed.

One way to frame an answer is to create a categorization of accreditation approaches, courses of action, and best practices within a standard set of management tools that can be structured into the common operations or duties that most managers carry out in many organizations. Some executives in higher education have compared the accreditation process to the sequential steps of the management process: planning, organizing, staffing, directing, and monitoring (Giacomelli 2002). In the mid-1930s, Luther Gulick wrote an introduction to *Papers on the Science of Administration* where he articulated these functions in the acronym POSDCORB (Gulick and Urwick 1937). For the purpose of managing accreditation within an educational institution, this includes the higher-level functions of:

- *Planning*—Formulating the accreditation strategy to allow the educational institution to exploit its core competencies, maximize opportunities, and meet the demands of the external environment, including accreditation agencies;
- *Organizing*—Aligning the organization's administrative, faculty governance, responsibility, and account structures with its strategy;
- *Staffing*—motivating and inspiring individuals to serve the needs of the organization, particularly with regard to accreditation strategy and planning. Topics will include recruiting, training, and directing their activities to achieve the stated goals;
- *Developing*—Creating a culture and a web of intraorganizational relationships that strengthens and maintains the core competencies and reinforces the accreditation planning, staffing, strategic goals, and organizational strengths;

and lower-level functions such as:

- *Controlling*—Monitoring and enforcing rules and procedures, encouraging productive and discouraging unproductive behavior, rewarding performance that achieves accreditation objectives and institutional improvement;

- *Operating*—Detailed planning and capacity utilization, scheduling of accreditation committees, and information collection;
- *Reporting*—Preparing and writing the accreditation self-study report to the accreditation agency, and responses to the visitation team's final report;
- *Budgeting*—Assessing and planning funding requirements for the various accreditation programs and objectives.

The POSDCORB approach is useful because it not only reflects a commitment to generic management concepts, it also satisfies the requirement that a classification system be mutually exclusive, contain very comprehensive clauses (Jones et al. 2001), and be clearly linked to various accreditation activities and best practices that will be uncovered. It is into this structure that many useful but practical, innovative yet achievable, unique yet not too uncommon examples of accreditation practices at colleges and universities will be detailed and explained in the next chapters of Part Three. Although these terms were listed at the bottom of figure 4 in Chapter 8, which examined the accreditation self-study, these examples will show that show that POSDCORB concepts can be applied to more than just institutional self-study processes. Although a majority of the examples will indeed be related to self-study activities, there will also be examples of how management concepts described here can bring together innovative, practical, interesting, and useful approaches to accreditation management that many colleges and universities have developed. It will then be up to other educational institutions to look at the POSDCORB principles as applied to accreditation management to adapt best practices, create an overall accreditation management strategy for their organization, and achieve the recognition and renewal that is sought.

Additionally, since the conventional functional paradigm of the POSDCORB approach can be viewed as a bit too strictly inward-looking (Jones et al. 2001), and since many new management approaches have evolved in recent years to include such topics such as ethnic diversity, globalization, quality improvement, organizational learning, and others, there will also be examples of leadership in the accreditation management and in the application of the Baldrige Award criteria described in Part Four of this book. A frequently contested question is whether leadership is a different behavior and occupation from management (Capowski 1994), and whether selected official leaders implement true leadership and others employ management techniques (Robbins 2000). John Kotter at Harvard University has stated that management actually entails handling complexity, and that excellent management results in stability and evenness by generating official tactics, conceiving organizational arrangements, and comparing results against the tactics (Kotter 1990a, 1990b). Kotter proposes that leaders create organizational direction by launching a vision for the institution's future and then gain support from people by conveying this vision effectively and inspiring them to surmount any obstacles. Leadership as well as effective accreditation management approaches are viewed as necessary for achieving organizational success. Therefore, examples of administrative leadership and knowledge management in accreditation by institutional leaders will also be illustrated later in this book, following the many examples of management in accreditation best practices at colleges and universities.

Innovative management techniques and engaging leadership approaches are particularly important for higher education because of the nature of the employee population and the distinctive organizational structure that has unfortunately been criticized of late, despite the many achievements of this unique industry (Birnbaum 1988; Henninger 1998). The distinguishing characteristic of this situation is the concurrency of administration and faculty governance command structures, in which different and numerous organizational decisions are deliberated and completed but occasionally not made, which often results in the continuation of mediocrity. There has been much emphasis lately on the significance of performance results in education—learning outcomes, in education parlance—and this has continued to be a critical issue for the general population, lawmakers, and accreditation agencies. In fact, the president of the CHEA), Judith Eaton, has stated that accreditation agencies have sought to restrict the linkage between value or worth and outcome results, and these labors have faced conflict in the higher education arena as a result of uneasiness and uncertainty (Eaton 1999). This conflict is likely to appear when college students are associated with regular consumers (as in commercial enterprises) and postsecondary education is treated like a business market. The academic culture has a long history and tradition that views colleges and universities as communities of intellectual growth and knowledge development, which is not easily connected with a result- and market-driven perspective. In addition, discomfort emerges when the quality discussion turns to outcomes because outcomes are usually attached only to student learning, and many institutions view this as part of their mission, along with research and service. Outcomes are also difficult to measure and not easily quantifiable, and many faculty are already involved with faculty development programs to increase student learning. Confusion can result when the talk is about quality of resources in some circles and quality of results in others. The level of resistance, discomfort, and confusion can be diminished by effectively examining the accreditation process, managing the important elements, and communicating the message to the stakeholders of the educational institution.

Accreditation management should consider these factors when formulating a complete institutional schema where the pursuit of or renewal of accreditation is harmonized with institutional organizational reviews, strategic planning, and distinctive organizational issues. Figure 5 demonstrates how accreditation management concepts fit between these elements and the general goals of achieving recognition by one or more accreditation agencies and systemic organizational development (Alstete 2004).

Colleges and universities have the specific task of making the accreditation practices a productive occurrence. The relatively new concept termed accreditation management is a principled strategy employed by postsecondary leaders to make accreditation a positive, enlightened, and educational community experience, and an effectively designed experience (or experiences, in case of multiple accreditations) instead of a response to an intervallic target as it is still perceived by many today. Over two decades ago, a book titled *Understanding Accreditation* anticipated several notions for making accreditation valuable, including making accreditation a top concern within educational institutions, appropriate budgetary support for accreditation functions, and managing instead of reacting to accreditation standard requirements (Young et al. 1983). Accreditation is seen by many educators and some individuals outside the education

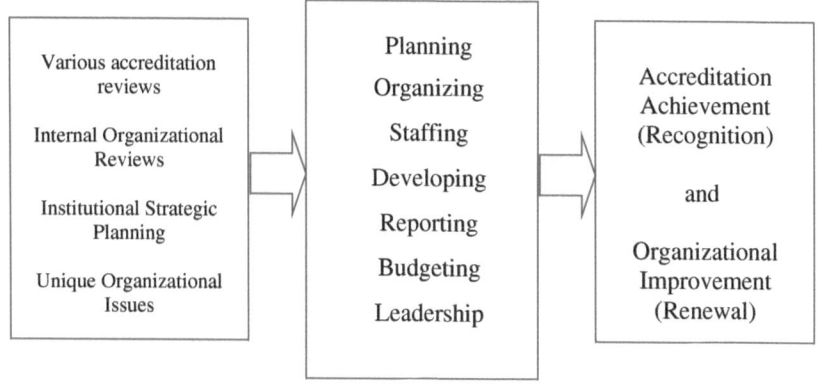

Figure 5 Accreditation Management and Institutional Goals

field as a difficult external influence for intermittent contention; for others, it is either a trivial irritation or a calamitous danger. Internal re-edification about accreditation significance and remuneration, alongside management of accreditation, can and does encourage institutions to attain recognition and revitalization less laboriously and more successfully than is conventionally the case. Without accreditation management, unique objectives, and effective processes for evaluating the outcomes, institutions may run the risk of having to allow visiting accreditors to intrusively impose goals, processes, and micromanagement on them. Accreditation management is not the solution to all of the challenges and difficulties related to accreditation, but it is undoubtedly better than not scheduling accreditation and enabling the college or university to be affected by assorted requirements from inside and outside forces.

CHAPTER 13

Planning Accreditation

Planning the accreditation management processes involves the arrangement and the procedural aspects of shaping how the organization, in this instance the postsecondary educational institution, will attain its goals (Cerbo 2003). This can be accomplished by analyzing, appraising, and choosing among the openings that are anticipated and that help the institution and people involved to be forward-looking. It is common for faculty, administrators, staff, and others to stay focused on the institution's day-to-day operations, challenges, and educational cycles. A sound preparation program improves decision synchronization by requiring the institution to look beyond the typical daily challenges to anticipate the challenges that may be encountered and the goals that can be attained, such as accreditation Arrangement and preparation are actually the principle management purposes, and the one that goes before all others as the basis for coordinating, persuading, and directing.

Generally stated, setting up a three-step method entails: (1) determining an organization's mission and goals; (2) formulating a strategy or strategies to achieve these goals; and (3) implementing the strategy (Jones and George 2003). An effective plan builds assurance for the organization's objectives, provides the college or university a sense of supervision and reason, organizes the roles and specialties, and controls the various people in the organization and internal sections. Strategic and operational tactics differ in their extent and time frame— strategic tactics apply to the entire organization over a wider time frame and operational tactics detail the plans of how the goals will be accomplished (Robbins 2000). Proper preparation not only gives leadership, it can also lessen the influence of transformation in the situation, reduce waste and idleness, and establish the principles that may be strongly required in highly decentralized, horizontal hierarchies that are relatively frequent in higher education. However, planning is not without condemnation. Some of the criticisms against it claim that it introduces too much inflexibility, which can stiffen individuals as well as academic departments into specific hardened objectives when the external surroundings are rapidly shifting; that it is not a substitution for instinct and invention (Mintzberg 1994); and that overplanning can center too much on beating the competition today instead of in the future (Hamel and Prahalad 1994). Also, planning can cause excellent organizations to become excessively attentive to their successful aspects and in so doing create the conditions that can facilitate failure (Miller 1993). It is important for accreditation managers in colleges and universities to remember these concepts when crafting accreditation management strategies as well as other operational and strategic plans for their institutions.

The related literature treats strategic planning and accreditation as distinct topics, but the two processes share numerous fundamentals (Barker and Smith 1998). These common fundamentals include an assessment of the institution's mission and goals, specific tactics to meet the goals, and an evaluation of how well the objectives were achieved. The self-study segment in a typical accreditation procedure can be quite long-lasting and requires considerable institutional resources. Normally, colleges and universities today begin the groundwork for the self-evaluation roughly two years prior to the peer-review team's scheduled visit to the campus. However, if the objectives of the accreditation strategy are aligned closely with the institution's strategic plans, then groundwork for the team visit can be achieved without directing all of the institution's labors, assets, and tactics into that 48-month time frame. Consequently, successful accreditation management should endeavor to position the accreditation process as a continuing procedure to foster organizational unanimity, improve utilization of assets, and, notably, guide the institution to complete the self-evaluation more capably and with greater significance.

The oversight agency for accreditation association, the CHEA, specifies that the accreditation procedure for postsecondary institutions is handled directly by the chief academic officer, often titled the vice president of academic affairs, provost, or dean (Eggers 2000). While individual faculty members are usually appointed to internal accreditation task forces or committees, and often to the volunteer peer-review teams, the overall direction and operative planning for accreditation is typically the job of the institutional administration. Colleges and universities should strive to coordinate their required regional and aspirant professional/specialized accreditation under the leadership of the chief academic officer, so that there is a single, inclusive cycle of organizational preparation and review. This combined and wide-reaching unitary accreditation process would in the end be less onerous on the institution and internal stakeholders, produce better accreditation-review quality, and encourage the participation of more individuals in the institution. A number of colleges and universities have already started to implement this innovative concept at their institutions, initially by requesting that their accreditation associations be flexible in their timetables, and subsequently by scheduling the self-study reports and visits onto a master calendar. This strategic accreditation management process can enable increased synchronization across the various specialized academic departments (such as communications, business, education, and health sciences) along with the required regional accreditations, to more effectively utilize databases, and improve the overall institutional asset and budgetary tactical planning.

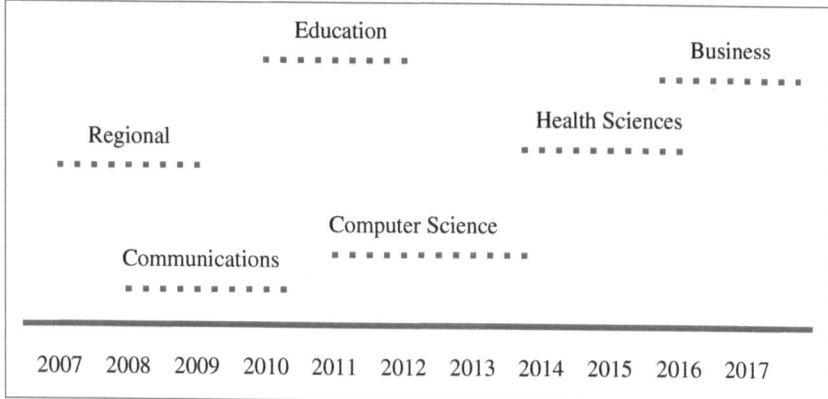

Figure 6 Unplanned Accreditation Time Schedule

A sample unplanned time schedule for several accreditation reviews at a typical college or university is illustrated in figure 6 (adapted from Alstete 2004), where the re-accreditation interlude between the required reviews may span over a standard ten-year time period. The usual two-year groundwork stage for each specialized or the one regional accreditation creates an ongoing situation where there is practically no time in a ten-year period at the institution when one or more academic accreditations are under review or in a preparation phase. Additionally, the college or university must also undertake strategic planning and visioning that will also coincide with these accreditations. These types of strategic planning are quite common today and require significant institutional effort, and often usually call for a two-year review for each five-year strategic institutional plan.

Pragmatically, it may not be feasible to coordinate all of the accreditation time schedules. Nevertheless many accreditation associations are usually able to be fairly accommodating, and if even just a few accreditors are willing to allow coordination, many of the abovementioned benefits can be attained, including significant budgetary reductions, improved results, and the reallocation of individuals and assets for other important institutional priorities. An improved and managed accreditation time schedule is illustrated in figure 7 (adapted from Alstete 2004) with a highly coordinated ten-year plan, along with the institution's strategic planning cycle shown in five-year periods. This accreditation management approach provides a philosophy for studying the college or university in coordination with the goals of the regional and specialized accreditation agencies, and better facilitates the utilization of institutional information that is currently available, as well as improving the system for continuous organizational outcomes assessment and managerial decision making. In regard to the overall strategic decision making of which accreditations to pursue, institutional leaders can conduct an environmental assessment to determine their strengths and weaknesses, and decide which specialized accreditations should be sought, reaffirmed, and possibly even ceased if it is not in the best interest of the long-term direction of the institution.

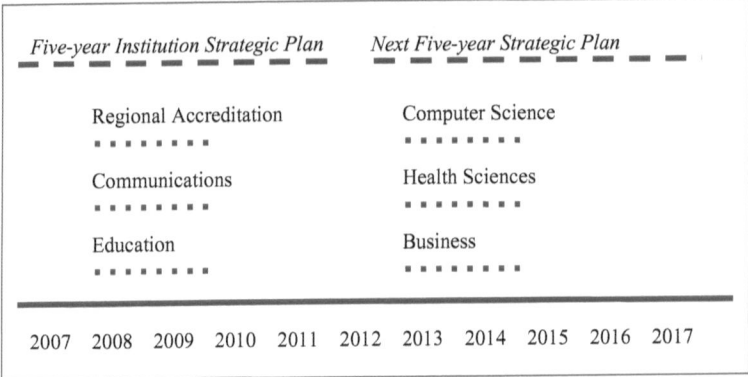

Figure 7 Planned Accreditation Time Schedule Linked to Strategic Plan

Even this sample illustration is fairly optimistic about effective accreditation management and should be viewed as a long-term goal. It does express the possibility of coordinating the numerous accreditation association schedules and requests that they adjust the previously scheduled time frames to harmonize with the institution's other needs. One minor complication to the accreditation time schedule planning is that a number of accreditation agencies have phases that are not in ten-year rotations, and therefore may not directly coincide if the accreditation association is unable to be adaptable in its reviews. Effective accreditation management strategies consider many alternative possibilities to solve challenges and make improvements, and preferably can arrange recurring yearly revisions for ongoing enhancement.

For more complete and successful institutional strategic planning at colleges and universities, a postsecondary institution should have a designated "chief planner," frequently the chancellor or president, and preferably involve a mixed collection of administrators, faculty members, and students to help create the institution's strategic plan (Cahoon and Pilon 1994). This broad-based team can be led by the chief planner in the area of accreditation management, and the team could also work properly as the steering committee for regional accreditation self-evaluations. The benefits include reducing startup time, helping to guarantee knowledgeable personnel, and facilitating access to important institutional information. An example to illustrate how closely the accreditation management approach can steer the self-evaluation for a regional accreditation study is exemplified in the College of St. Scholastica in Duluth, Minnesota. This example will be briefly examined in order to show and discuss this concept. Importantly, this institution's strategic planning process was the managing construct for the accreditation self-evaluation procedure, and the strategic plan was updated annually through the late 1990s. The strategic planning team that conducted the annual updates was made up of the college president and the senior administrative cabinet members, which included the chair of the faculty assembly, the academic deans, the director of public relations, and various student representatives.

This plan, which was updated each year, was titled *Scholastica Direction* (41–42) and contained four sections:

1. The purpose, process, planning cycle, and overview of the planning document;
2. A review of the mission statement, commentary on the mission statement, and "Mission Statement – Lived Experience," listing the college activities that put flesh on the more idealistic assertions of the mission statement; long-range planning assumption describing expected external future conditions over which the College has no control but that will have an impact on the College; institutional characteristics of the College, including societal context, relationships with external agencies, leadership/management, programs, students, student services, staffing/development, physical facilities, equipment, and fiscal resources; a listing of the major strengths and weaknesses of the College;
3. Five-year institutional goals with selected objectives: these include one goal directly tied to enhancing understanding of the College's mission, a series of operational goals, and a goal focused on capital development needs; these goals are set in response to the weaknesses identified in the previous section of the plan;
4. An appendix that usually contains illustrative graphs, particularly supporting statements in the second section. (Cahoon and Pilon 1994, used with permission.)

The Plenary Planning Commission (PPC) at the College of St. Scholastica held meetings throughout each academic year, gathering perceptions and information from the college stakeholders and community, evaluating and changing various sections of the previous year's plan, and informing the college community of the continuous planning effort to solidify their support. Additionally, as is commonly known to college leaders who are experienced with regional accreditation needs, there is a fundamental link in the relationship between the college's strategic planning process and the groundwork of the self-study. The design for the self-evaluation at St. Scholastica included the intent, process, and timelines and was written in the first section of the college's yearly planning manuscript document. The self-evaluation also generally included a reaction to the previous regional accreditation report, and any important changes that may have occurred since the last peer-evaluation accreditation team visit. Subsequently in the next (second) segment of the college planning document, supplementary topics associated with the self-evaluation were included, such as an assessment of the college's mission and goals, regulation of institutional resources, and outcomes assessment procedures. The final accreditation self-study report turned out to be a more in-depth evaluation than the college's annual planning endeavors; nevertheless, the self-study committees that were assigned with specific sections of the report had a much less troublesome beginning than they would have had if no strategic planning system were being implemented.

One interesting aspect of this example at the College of St. Scholastica is that the institution changed its senior leadership in the late 1990s and disbanded the PPC in favor of an outside strategic planning adviser. Even though the subsequent periodic ten-year accreditation evaluation was not a failure, the regional accreditation evaluation suggested improved concentration and effort on institutional strategic planning. Following this period, a reformation of the internal institutional strategic planning team is happening as well as a revisit to the systematic annual updating of the strategic plan (Cahoon 2003). Nearly all accreditation association standards today incorporate assessment of academic achievement and assurance of student learning as a significant portion of their requirements for initial and renewal of accreditation approval. The planning strategies by the institution must include these plans and document ongoing efforts from the very start, so that a full demonstration of institutional support can be articulated for academic, and most nonacademic, support programs. If the college or university has an ongoing annual strategic planning process like that of the College of St. Scholastica, then accreditation and assessment preparation is also greatly enhanced by a treasure of information that documents developments of organizational innovation and performance within the institution. Accreditation management strategies that include outcomes assessment planning will be examined in more detail in a later chapter.

When colleges or universities go through an accreditation self-study process, individuals involved with the range of discrete responsibilities rather than with a planned accreditation management strategy of vital assessments that are linked to the strategic plan of the institution may become confused. In particular, the dues and requirements to accomplish the accreditation, not to mention strategic institutional planning, can appear intimidating (Giacomelli 2002). Despite this challenge, the situation can be addressed by understanding from the beginning that each responsibility or endeavor in the seemingly disconnected process is an integral part of the institution's plan, and that the accreditation management process and the overarching goals (accreditation recognition and institutional renewal) are extremely valuable. A good accreditation management strategy is customized for each institution, and there is no single design that will suit all kinds of college and university missions, organizational cultures, and learning environments, on which the management approach should be based. However, there are some opening inquiries that should be studied as the initial step in the planning process, since the resolutions to these questions will have influential effects on the accreditation management and the strategic plans for the institution. For instance, it should be asked whether the self-evaluation is seen as a positive prospect for improvement or a threat for losing resources. Is the self-study at the institution continuing and vibrant or has the college or university experienced great difficulties in accreditation? Are the managerial decision-making procedures centralized or decentralized? (Giacomelli 2002). Additionally, the initial setup of the plan should also show whether the institution has a wealth of usable information, is lacking in data, and what ongoing internal evaluation procedures are currently being done that can be used as a foundation for improvement.

In designing an effective accreditation management plan, an appropriate work configuration is important to "support the key actions of the self-study, such as communicating, documenting, analyzing, interpreting, synthesizing, evaluating, and prioritizing." (Giacomelli 2002, p. 284). In order to achieve this, the accreditation plan should specify who at the institution will perform the duties (the plan), explain the methods of work (the process), lay out the time schedule, and list the final product contents. The achievement of the management accreditation planning process will be based on the philosophy that steers its design. In another example at Robert Morris College in Chicago, Illinois, the model of planning, process, and product was utilized in a comprehensive self-evaluation study for their regional accreditation renewal with the NCA. The product was a self-study result that accomplished many goals of the accreditation pursuit, contributed to important institutional development, and enhanced educational and organizational quality. Strategic planning that included the plan, process, and product was employed regularly during the ten-year interval that led up to the recently mandated periodic evaluation. The officials at Robert Morris College used this approach to commence initiatives such as the expansion of new campus locations, a higher offering of degree levels, and the addition of degree programs. Besides the enhancements at the institution, the results from the regional accrediting association produced the most important achievement of continued accreditation with the following comprehensive review scheduled for ten years later, along with concurrent approval of three institution update requests and no compulsory accreditation update reports or planned visits for ten years.

The education and lessons obtained from examples of other colleges and universities that connect these important procedures can offer helpful models for others in the critical combination of self-study and strategic planning efforts. In another example, Estrella Mountain Community College, which is a fairly newly established college in Arizona, showed innovation by integrating their strategic planning efforts as a central and continuous part of the culture within the campus community (Goodman and Willekens 2001). The planning at Estrella has gone through several discrete stages since the institution was founded in 1988, with the first phase focusing on six planning aims:

- Collaborative strategic planning
- Educational responsibilities to the area communities
- Comprehensive instructional programs and flexible approaches to instructional delivery
- Partnerships and collaboration
- Integration of information technologies across the curriculum
- Strong linkages with public schools

The second phase of the institutional strategic planning concentrates on departmental planning. Each academic and administrative department is expected to prepare a plan for their division that contains clear connections to the strategic plan of the college. The workings of each planning effort include regular items such as the department mission, goal, or purpose that is connected to the strategic plan, summary of resources, and a continuous improvement plan. The finished plans from the various departments are used as an important foundation for divisional budget requests and overall strategic institutional budget planning. This connection of important functions and senior administrator support, especially in the budget oversight and approval, is crucial for the achievement of any pioneering strategy in planning and accreditation management that diverges from the traditional operating methods that are frequently and commonly used at most colleges and universities. The tertiary step of the planning at Estrella Mountain Community College included the problems previously addressed during its primary accreditation self-evaluation, and were integrated into the strategic plan. The Leadership Council of the institution directed the creation of the college's strategic plan along with the other important plans, and anticipated that the strategic plan would direct all arrangement, guidance, and institutional resource allotment.

The vital connection between a regional (and sometimes specialized) accreditation self-evaluation project and institutional strategic planning too often goes unnoticed by colleges and universities. This linkage can be a way to facilitate constructive feedback in their strategic planning efforts for renewal and discovery of new or extended strategic concerns. As discussed earlier, regarding the self-evaluation procedure, a SWOT (Strengths, Weaknesses, Opportunities, and Threats) analysis can also perform a vital role in a properly managed accreditation self-evaluation. Estrella Mountain Community College's leadership team also utilized a SWOT analysis as an element of the connection between their regional accrediting agency, the NCA, and the strategic planning function of the institution. The linkage of the planning with the NCA's five criteria for accreditation is illustrated in table 3 below, and reveals the possible influence on the strategic planning process:

The self-study review conducted in 1996 by Estrella Mountain Community College had a considerable effect on the institution's process of strategic planning, and produced an outcome that included a set of six institutional challenges and 36 self-identified proposals for improving the institution. After reporting the final decision of the self-study conducted by the accreditation association to the college stakeholders, the college leadership then also connected these questions to the institution's strategic planning procedures. Table 4 illustrates the connections between the college's acknowledged challenge areas from the accreditation self-evaluation and the strategic directions from the institution that is a primary outcome:

The accreditation self-study in 1996 also had an influence on various other planning areas, in addition to supplying direction to the updated strategic plan at Estrella. Some of the changes that Estrella Mountain enacted based on self-evaluation results include modernizing the vision of the college; reexamining the mission and goals; development of core values; improving institutional effectiveness and processes of academic assessment; and reforming and clarifying all institutional planning methods. This final enhancement to the Estrella Mountain planning process exemplifies effective integration of strategic planning with the management of accreditation at the institution. The college example discussed here illustrates various lessons that are beneficial for other institutions to learn as well—in particular, the need for expansion of ongoing organizational planning and evaluation rotation, and the creation of durable connections to the managerial decision-making and budgetary allotment procedures. In addition, the importance of linkage to assessment strategies should not be trivialized, as well as the strong need for collaboration in a successful planning process. Planning is not as successful as it could be if it is not performed with integrated organizational collaboration by the college or university, with an effective process that increases individual participation results for all the people at the college who are involved to gain buy-in and knowledge.

Table 3 Connection of Regional Accreditation Criteria and College Strategic Planning

NCA Criteria	Link to Strategic Planning	Potential Impact on Strategic Planning
Criterion One: The institution has clear and publicly stated purposes consistent with its mission and appropriate to an institution of higher education.	-Mission -Purposes (Goals) -Strategy	Serves as a thorough review, and may result in changes to mission and purposes (goals). In extreme cases, changes to mission and purposes may cause the institution to reevaluate the programs and services offered to its public.
Criterion Two: The institution has effectively organized the human, financial, and physical resources necessary to accomplish its purposes	-SWOT -Budget Planning -Decision-Making Processes -Strategy	Identifies the institution's strengths and weaknesses related to all forms of human, physical, and financial resources. Provides an evaluation of the resource allocation process that may result in changes to budgeting and decision-making systems.
Criterion Three: The institution is accomplishing its educational and other purposes.	-SWOT -Budget Planning -Decision Making Process -Strategy -Institutional Effectiveness and Assessment	Serves as an internal scan that can identify the institution's strengths and weaknesses related to achievement of the college's mission and purposes. Identifies program and services that may need special attention in the planning process and require additional resource investment.
Criterion Four: The institution can continue to accomplish its purposes and strengthen its educational effectiveness	-SWOT -Budget Planning -Decision Making Process -Strategy	Identifies strategic issues that may challenge the institution over time. Provides an evaluation of the planning, budgeting, and decision-makin processes that can result in improvement to program and service delivery.
Criterion Five: The institution demonstrates integrity in its practices and relationships.	-SWOT -Values	Serves as a check on the institution's integrity and values system, may result in changes to an institution's values and/or mission.

Source: Goodman and Willekens, "Self-Study and Strategic Planning," p. 294, used with permission.

Using a Focused Approach in Accreditation

Many institutional leaders and faculty members may not know that when planning for accreditation management and institutional strategic planning, colleges and universities usually have the opportunity to select from various types of accreditation reviews. Most of the regional and many specialized accrediting agencies offer several types of accreditation approaches. These vary in language, with designations such as "comprehensive," "comprehensive with emphasis," "focused," or "special emphasis."

Table 4 Linkage between Self-Study Challenge Areas and Strategic Planning

1996 Self-Study Challenge Areas	Current EMCC Strategic Directions
Institutional Planning	**Planning and Charting Our Future** *Estrella Mountain must fully implement a system of planning and assess progress toward its mission.*
Growth and Development	**Growing and Expanding** *Estrella Mountain must be proactive in meeting the needs of a growing West Valley population.*
Community Involvement	**Creating Partnerships** *Estrella Mountain must continue to engage in partnerships activities that advance the mission of the College.*
Organizational Governance	**Investing in People** *Estrella Mountain must continue to develop and invest in systems that support becoming a quality-driven institution.*
Strategies for Information Technologies	**Integrating Information Technologies** *Estrella Mountain must continue to invest in technologies to support teaching and learning the development of new delivery systems.*

Source: Goodman and Willekens, "Self-Study and Strategic Planning," p. 295, used with permission.

There are also innovative types that are connected to the Baldrige criteria or a separate quality initiative such as the Academic Quality Improvement Project (MSACHE 2002, p. 669; NCATE 2003, p. 670; AQIP 2002, p. 668). Each kind of accreditation model has certain advantages and disadvantages to consider in the planning process. In selecting the method or approach for accreditation at an institution, college or university academic administrators who make this decision should think about the potential of their institution, obstacles to surmount, and the various advantages that could be obtained by implementing each of the various models.

One of the most popular self-study methods is the comprehensive self-study approach, which permits a college or university to evaluate every feature of the institution's programs and offerings, administrative and support organization, resources, and student-learning outcomes in relation to the mission and goals of the institution (MSACHE 2002). A typical widespread comprehensive accreditation self-study evaluation normally starts with a thorough reconsideration of the institution's stated mission, goals, and objectives. This first part of the review sets the basis for gathering data, conducting examination of the information, arranging priorities, and determining suggestions for modifications and development in the college or university. The planned arrangement for a comprehensive accreditation review will contain a precise investigation into each section of the institution, though a number of areas might be chosen for even more thorough scrutiny.

The critiques stated in the first part of this book involve observations by many people that standard postsecondary accreditation evaluations, which are usually the comprehensive type, are not a considerable hurdle for most colleges and universities today. Educational leaders should therefore strive to make the postsecondary accreditation procedure at their institutions an authentic assessment and enhancement process that is an example to be respected so that the potential unease of the public, legislatures, as well as other external stakeholders is addressed. One option is to implement a special-emphasis or focused self-study as an alternative to the comprehensive option for previously accredited, reputable, well-functioning colleges and universities that are ready to give attention to a chosen set of significant issues and topics that will play an important role in enhancement and educational distinction (Winona 2003). It should be emphasized that focused reviews are not usually available to educational institutions that are pursuing their initial accreditation with an association. Since the standard or conventional comprehensive evaluations actually contain many advantages, these special-emphasis accreditation self-studies should be considered only after a thorough analysis and determination by institutional academic leaders and senior administrators. While a comprehensive accreditation review is an outstanding occasion for a college or university to present for inspection some important concerns without specifying particular programs, academic departments, or problematic issues, this type of evaluation does frequently invite wider involvement from institutional constituencies. An advantage of a focused evaluation accreditation process is that it can facilitate renewal for some postsecondary institutions if they take advantage of this opportunity to construct their self-evaluation procedure around a selected and relatively small number of thoughtfully chosen vital areas that they wish to enhance or excel in. Certain colleges or universities may wish to examine an institutional topic that strongly requires attention, such as alumni affairs, academic services, new degree program development, or student affairs. This type of specific accreditation approach can be a large help to the many institutions today where the regular comprehensive accreditation review is not a challenging hurdle and all of the internal effort involved provides negligible benefits and return on resource investment.

In the strategic and accreditation management planning stages, leaders of educational institutions can ascertain whether they should utilize the focused review option by considering whether the institution has been regionally accredited for many years, including at least one ten-year round between two regular comprehensive evaluations. The institution should also: exhibit that it has sufficiently developed internal assessment programs for evaluation and sufficient institutional research to supply the appropriate information to sustain the college's or university's assertion that it exceeds the criteria for accreditation by the association; prepare written documentation that verifies that there is a solid agreement among important institution stakeholder groups that the topics of special emphasis are suitable, well timed, and are among the key decisive topics or issues that are facing the institution; provide a self-evaluation arrangement that exhibits that the progress on the subjects of focused review will connect with a large proportion of the institution's community; and give evidence about the institution's willingness to act quickly in response and confidently to the accreditation review suggestions that are the product from examining the focused areas and the institution's readiness to be evaluated on its deployment of the accreditation results of this special-emphasis self-study (Winona 2003).

An accreditation internal steering committee at postsecondary institutions that are considering a self-study procedure constructed around a specific set of special-emphasis topics should speak with the accreditation association staff liaison in the first part of the accreditation management planning process. The college or university should send a well-prepared plan for self-study that justifies the chosen focused areas and that specifies the method of evaluation to examine these issues in depth and how the institution will validate that it continues to meet the General Institutional Requirements (GIR)s and Criteria. The accreditation commission personnel must review and concur with the suitability of the focused review option for the institution. The agency personnel then write an official approval that outlines the accreditation commission's agreement to allow a special-emphasis review. These approval documents supplement the peer-evaluation committee invitations and all accreditation agency materials that are utilized throughout the review procedure.

As stated earlier, a special-emphasis or focused self-evaluation procedure can be positioned as an occasion to reassess and possibly improve the college or university mission; to examine student enrollment trends in an institution; to begin a more involved and thorough procedure for measuring student learning; to evaluate the effect of a revised organizational governance structure; to create a successful long-range strategic plan; and to measure and adjust the strategic plan. Some additional possibilities might include program assessment, student service-learning, cultural or intellectual diversity, review of general education, postgraduate curricula, undergraduate programs, student and faculty information technology needs, faculty and institutional research, renewal procedures for strategic planning, and student development. A focused accreditation self-evaluation should lead to solid and honest development and improvement in the topics covered. This type of self-study procedure strongly encourages and even requires that proposals and results that emerge be given full attention and receive important feedback from the senior administration leadership.

Like the traditional standard comprehensive accreditation self-evaluation studies, the focused or special-emphasis self-study report normally contains a table of contents, an introduction section, a main body of content, and a summary section. However, there is an additional opening section in the special-emphasis self-study report that explains why the college or university undertook a focused accreditation self-evaluation approach. The beginning section of the report body usually provides concrete verification that the fundamental accreditation standards for the accreditation association have been achieved. A large part of this substantiation can be supplied by explaining information about the institution that is readily available. The next section of the main body should be dedicated to the special-emphasis topic or several emphasis areas that are being examined. The concluding section of the report should contain a summary section that presents the college's or university's reasoning as to why it should receive reaffirmation of accreditation for satisfying the standards and how it successfully utilized the special-emphasis self-evaluation as a valuable method to enhance institutional progress.

During the initial stage of planning and preparing the focused topic review that the peer-review team will study when it visits the institution under evaluation, the college or university should ensure that important institutional stakeholders, such as the president or chancellor, understand the meaning of the topics of special emphasis and display ownership of the results attained from the accreditation self-evaluation. The special-emphasis plan ought to anticipate that greater participation will be needed from those individuals and units that the peer-review committee interviews will be directly concerned with. These people should be alerted that the assessment of the topics of special emphasis would require them to have an unambiguous and clear awareness that they had the occasion to offer contributions to the focused accreditation process. Additionally, the plan should foresee that the most important concerns confronting the institution will be included in the focused topics and that there will be no unanticipated key problems that the self-evaluation did not deal with. Since the institution has devoted itself to embarking on a special-emphasis self-evaluation, the visitation committee may advocate additional monitoring by the agency (such as required additional progress or interim reports) than would normally be required in a traditional comprehensive self-study. The personnel or managers at the accreditation association usually propose the members of the peer-review visitation team in consideration of focused topics that were explained in the written proposal by the college or university, and this element should also be thought of in the development of the plan. The special-emphasis accreditation evaluation committee for a focused visit is sometimes bigger than an accreditation team for a traditional comprehensive evaluation review team.

There is, as a minimum, one regional accrediting association in the United States that advocates support of trialing the newer review procedures for institutions with specific stipulations. These specifications include that the trial take place only with respected and thriving colleges or universities (those that are not high-risk institutions); that specially trained or oriented personnel will be assigned to the review committees involved in these experimental evaluations to help guarantee that they can carry out the responsibilities delegated to them; that an official assessment of the specialized procedure will be at the heart of each trial review; and that an experimental evaluation will not harm the institutions (Winona 2003). Additionally, the accreditation association will make certain that the peer-reviewers from other institutions are not held liable, that the cost arrangement will pay for true committee expenses as well as a general administrative fee, and, perhaps most notable to the accreditation management strategy, that a college or university interested in creating and carrying out an experimental review procedure should work together with the agency staff liaison in the first part of its arrangements and planning.

The focused or special-emphasis accreditation review is still considered fairly recent and is not yet generally known to a lot of senior administrators in higher education or extensively used. When contemplating this type of accreditation management and planning, various questions should be thought about, such as: What precisely is involved in a focused accreditation evaluation? How does a special emphasis differ from the standard comprehensive self-study? What reasoning will be utilized for adopting this innovative approach for accreditation self-study? How do colleges and universities perform the focused type of self-evaluation (Hutchinson 1994). One institution that planned and implemented a regional accreditation self-evaluation using a focused or special emphasis is Bowling Green State University (BGSU), a Carnegie Category I doctoral-granting university enrolling nearly 20,000 students in Ohio. At the institution, the special-emphasis approach is described as "more focused" than a traditional comprehensive study by the associate vice president for academic affairs, Peter M. Hutchinson (Hutchinson 1994, p. 79). A more precise description of a focused accreditation review is to see this planning policy as more "thematic" where the educational institution defines various detailed viewpoints through which to study how the college or university has improved since the previous accreditation self-evaluation and what else it can do to advance. The special-emphasis accreditation strategy is seen by some institution leaders as more challenging than planning and conducting a conventional comprehensive self-evaluation. For instance, all academic departments and support areas of the college or university must carry out a separate self-evaluation, comparable to the method or procedure these departments would be required to complete in a traditional comprehensive study. The difference in this approach is that the results produced from these individual departmental or nonacademic support unit self-studies can then create the basis for the required comprehensive portion of the special-emphasis report. This was mentioned earlier as the first section of a typical focused evaluation. In addition, these reports can provide much of the needed material for the focused sections of the final special-emphasis report that is submitted to the accreditation agency if the departments are guided properly on the accreditation approach.

Managers of accreditation at colleges and universities may see this focused or special-emphasis strategy as inherently more unclear and containing greater increased risk than a customary comprehensive accreditation review. However not all institutions hold this perception, and the accreditation managers and institutional leaders at BGSU did not think that this negative attitude was widespread during BGSU's self-evaluation. One significant benefit of the thematic approach is that the rewards to the institution are better, and this can validate the increased risk that is sometimes seen by some stakeholders involved in the endeavor. The focused or special emphasis obliges that the college or university intensely study those concerns that are most critical, imperative, and required, and it provides the opportunity of considerably better results from an accreditation project than a traditional comprehensive evaluation. On the other hand, a focused accreditation strategy may not be the best option in an accreditation management plan for every postsecondary institution that needs to go through accreditation. Educational institutions that are in a relatively early phase of their evaluation, or are pursuing their initial regional accreditation, should proceed with a conventional comprehensive review because such institutions must first identify and acknowledge their primary general level of organizational quality before seeking to explore detailed problems in a focused accreditation review. Those institutions that need to go through continuing reaffirmation of accreditation and have an established lengthy record of regional accreditation could undoubtedly take advantage of this distinctive, significant, and still somewhat novel approach. It is understandable that many or probably most college and university leaders, and particularly faculty members and middle-level administrators have yet to discover and learn about the range of accreditation methods that are frequently available to utilize at their institution. Educating these colleagues about the alternatives should be planned as part of the accreditation management strategy, specifically in the initial phase of an accreditation plan, and in the longer-range institutional strategic plans that are critical to success today. A focused or special-emphasis accreditation method is most suitable for colleges and universities that have very specific objectives and issues that need to be examined, that strive to account for their current successes (such as the BGSU and other postsecondary institutions with a comparatively lengthy record of regional accreditation), and that may have an important challenge or challenges to address with external guidance.

In planning a focused topic accreditation evaluation, the experience and results of a university such as BGSU can demonstrate to other institutions how important it is to contemplate issues such as the planning, incorporation, synchronization, and possibility for replication within the institution's traditions and behavior patterns. Even though these factors may seem apparent, they are central in the accreditation management process, and a proper means for managing these matters can be effected by the institution through the selection of suitable employees, external accreditation review team choices, internal accreditation steering committee assignments, and properly appointed support staff (Hutchinson 1994). An important topic that BGSU leaders discovered in their implementation of the focused or special-emphasis review for their regional accreditation was the makeup of the peer-evaluation team visit. The external committee visit to the campus should be arranged specifically for a special-emphasis review, because the accreditation team members who are deployed to this university did not appear to be very accustomed to the special features of a special-emphasis accreditation strategy which can cause this evaluation to be fundamentally different from a traditional comprehensive evaluation. Particularly, the external review team plan was arranged quite similarly to a comprehensive evaluation, and the peer-evaluation team members did not meet with the university's special-emphasis internal committees that performed much of their self-study. This created some misunderstanding and even some resentment among the special emphasis team members who truly believed that they had learned some unique institutional perspectives which were not properly reviewed by the accreditation review team. The accreditation association associate, as well as the university, probably should have planned and managed this focus review better in this regard so that they could have benefited more fully from this special emphasis review and helped the institution resolve its distinctive concerns.

Now that some lessons have been examined as to the potential negative consequences of poorly planning a special-emphasis strategy, an example of a positive result with a focused accreditation method will be illustrated with the recent regional accreditation review at the University of Michigan–Ann Arbor by their regional association, the Commission on Institutions of Higher Education of the NCASC(UMICH 2000). The university prepared for a special-emphasis visit for its required self-evaluation with a focus on the topic of "New Openings for the Research University: Advancing Collaborative, Integrative, and Interdisciplinary Research and Learning" (p. 3). The accreditation review on-site visit to Ann Arbor was performed as a two-part activity, with the 12-member accreditation team being divided into two subteams. One subteam of four external peer-review members focused on the university's conformity with the GIR and the Criteria of the association for accreditation, which all institutions must fulfill to receive NCA accreditation. While the other subteam (consisting of eight peer-reviewers) concentrated all of their efforts on the specified special-emphasis project, this two-team accreditation review strategy permitted an illuminating strategy to analyze the university's integrity to external reviewers in a representative, yet somewhat traditional, and positive viewpoint, while also elucidating new organizational strategies or approaches to the powerful external peer-reviewers from the association. The two subcommittee teams visited Ann Arbor and evaluated the university with positive results, including recommendations to reaffirm the regional accreditation and to schedule the subsequent accreditation review in ten years. The innovative accreditation approach applied at Ann Arbor demonstrates how a well-respected top research university can communicate its intentions to internal as well as outside stakeholders using methodology that is positive to the organization, with a procedure that has linear fundamentals of management such as planning, organizing, supervising, controlling, and reporting, yet deciphers or translates the accreditation strategy uniquely with a focused approach on one or more important topic.

CHAPTER 14

Organizational and Staffing Issues in Managing Accreditation

A very important aspect of effectively managing accreditation is choosing the best personnel, motivating them to join the accreditation effort, and leading these essential individuals who are involved on teams in the accreditation process. Several fundamental issues in effective staffing in accreditation management are illustrated in the case of Eastern Michigan University (EMU), which completed a successful regional self-evaluation process that resulted in the association's decision to continue the accreditation for ten years until the next comprehensive evaluation (Bennion et al. 2002). Even though senior administrators and academic managers at the university understand that there are a variety of reasons for this level of achievement in the self-study endeavor, they have stated that a large part of their accomplishment can be credited to the successful work of the institution's internal 28-person self-study steering committee. The assignment of members to an accreditation steering committee is a vitally important decision that an accreditation management team must make when organizing and making arrangements for the accreditation evaluation. Consequently, the senior administrators at EMU suggested to others that the procedure for personnel selection should not be engaged too quickly. The assignment procedure at their university actually continued for a complete semester as the dean's council conferred with the accreditation self-study coordinator. It wanted to choose individual members for the accreditation team who not only symbolized various departments and areas at EMU, but also were racially diverse with excellent thoughts to offer the group, and who were willing to perform the challenging work needed for writing an accreditation self-evaluation study. A list of committee members shows the titles of various committee members and lists the names of those who appointed them. The provost appointed a majority of the committee members, and other appointments were made by various vice presidents and senior administrators. The accreditation steering committee contained a selection of university administrators and faculty members, including department chairs, academic deans, student support unit managers, and two students.

A major issue for the self-study coordinator at EMU that involved the staff was to create buy-in or active engagement, and then to maintain the committee members' participation throughout the self-study procedure. The self-study coordinator asked for members' ideas at committee meetings, requested that each member participate on one or more of the various accreditation self-evaluation report chapter-writing subcommittees, managed to keep the committee genuinely involved in reading and editing the many drafts that were created, arranged attendance at public meetings on the report, and reported back to the accreditation committee about any problems or questions that were elicited by the university community. Additionally, the self-study coordinator wrote and distributed a monthly newsletter about the accreditation process at the university titled *NCA Self-Study Matters,* and this communication vehicle acknowledged the efforts of the hard-working committee members and also included articles written by some members. Even though there were no monetary incentives available to give to the committee members who served on the accreditation committee, some inexpensive yet emblematic rewards were given, such as team lunches, small pins, and even some nice tote bags for all committee members. These seemingly innocuous and even clichéd items are actually appreciated by people in academe, who are often underpaid and in need of additional recognition. Other colleges and universities have also published accreditation newsletters for staff and faculty members, and a sample list of accreditation newsletter topics is detailed in Appendix D. Once again, it is good management and communication efforts such as these that can create institutional stakeholder support and form part of an overall effective strategy for managing accreditation. As mentioned earlier, a research study on the academic dean's role in change, specifically the execution of new professional or specialized accrediting standards, was performed by interviewing faculty at four professionally accredited comprehensive universities in 1996 (Henninger 1998). The results of this research survey revealed that university leaders can induce organizational change, but only to a restricted degree. Faculty members in particular are deeply prejudiced by their academic or professional disciplinary communities as well as the local institution's expectations and limitations. For that reason, the accreditation management strategy should plan for these issues and potential problems, and specify the alternative and extra communication techniques that can be used to better surmount the obstacles, lead the faculty and staff on the committee, and positively influence the organizational culture at the college or university so that the accreditation(s) is achieved and the institution is renewed.

An accreditation self-evaluation procedure can be, and often is, perceived as a constructive test for individual faculty and staff members, as well as institutions who are seeking accreditation approval and institutional rejuvenation (Taylor 2002). An important aspect of the staffing management that college and university leaders and particularly the accreditation self-study coordinators ought to pay close attention to is assisting and guiding individual stress reduction for all members of the institution's steering committee. Even though most personnel in postsecondary institutions are typically very knowledgeable about the significance of the accreditation review procedure, the weight of the decision, and the penalty of not achieving approval, many individuals find the overall accreditation work to be somewhat stressful. This can happen even if the college or university had successful experiences in the past with regional or specialized accreditations, because some people may have personal doubts about their professional competence or any other previous experience in the report section for which they are now coordinating the writing in the accreditation self-study. The institution's managers who are directing and formulating the accreditation management plan may wish to solicit suggestions from the institution's staff counseling office, faculty members in counseling or psychology, or even the student guidance personnel, who are usually employed at these institutions and willing to help. Individuals with suitable backgrounds such as these may be appointed to the self-study steering committee, either deliberately or not, and their abilities can be leveraged for the project. These management strategies are part of an overall internally guided accreditation process to make the review procedure as effective as possible. A portion of the suggestions may involve recommendations about the method of managerial communication by the institution's leaders concerning the accreditation objectives, reduction of the self-evaluation-related stresses, and a synchronized program of relaxation methods for team members (including exercise, nutritional guidance, and team-building training) that produces synergy and decreases stress.

Another challenge in managing accreditation processes, and many other processes in postsecondary institutions, is the variability in decision-making styles, time urgency, main concerns, and differences among important stakeholders of the two very different faculty and administrative groups (Martin et al. 2001). The leadership and senior administrative philosophy of postsecondary institutions today are largely differentiated by an aspiration for effectiveness, capability, value, high output, and accountability, while long-established faculty cultures have been portrayed as promoting peer-review, collegial and mutually respectful relations, autonomy, curricula dominance, and a particular ranking of academic disciplines, among other things (Clark 1989, 1963). The disparity among the faculty and administration cultures in colleges and universities can create great challenges in an accreditation management procedure because it can hinder the effectiveness of communication and obstruct organizational change. Martin et al. (2001) state that these cultural obstacles to the attainment of clarity and control in accreditation reviews can be conquered by:

1. Instilling a discipline of reflection and a culture of evidence that supports honest discussions about the current condition of the institution;
2. Fostering new patterns of conversation and interaction that encourage and support the involvement of a broad spectrum of the campus community in defining the future and assessing institutional progress toward shared goals:
3. Promoting genuine conversation about difficult and controversial subjects as a way to disperse power and leadership throughout the organization;
4. Advancing a philosophy of experimentation, assessment, and management of reasonable risks associated with change;
5. Providing open access to meaningful information about the condition and resource base of the institution; and
6. Approaching planning as a scholarly endeavor that creates a research-based foundation for action. (p. 113)

The advancement of an institutional culture that mirrors these principles can help facilitate an effective accreditation management procedure triggered by an important accreditation self-evaluation. The outcome can be an improvement of the institution that is based on historical assets of both the conventional faculty and administrative behaviors which are a part of many colleges and universities. The cultures and inherent natures of both of these groups lead both to pursue acknowledgment and revitalization of the institution, however with somewhat different perspectives and objectives. Nevertheless, an effective accreditation management plan and strategy that understands and strategizes for institutional culture growth can assist both groups to attain the objectives that have been promoted and cultivated by the accreditation management plans and the institutional leaders who have permitted and encouraged them.

Other colleges and universities have also found that comprehending and improving the campus group behaviors is a fundamental part of achievement in the management of accreditation processes. A good example of this can be seen at North Dakota State College of Science (NDSCS), where institutional leaders discovered that employing a decisive management-planning procedure that occupied a greater part of the individuals at the institution and obtaining their support is key to the success of an accreditation endeavor such as a regional accreditation self-evaluation (Dohman and Link 2001). Initially the senior administrators at the institution should be prepared to robustly compel the accreditation strategies and become, if the leaders are not currently, very knowledgeable about the formal accreditation standards and informal expectations of the accreditation association. NDSCS was privileged to have an institution chief executive officer (college president) who was then currently serving as a regional accreditation agency consultant and peer-reviewer, and held a superb working understanding of the objectives and stated goals of the regional accreditation organization. If the institution's chief executive or other senior leaders require additional education and updating about the current accreditation association standards, then participating at the annual accreditation association meetings, attending separate accreditation seminars, visiting with institutional peers who recently finished a success accreditation endeavor, and communicating with the regional association's personnel liaison are just some of the techniques and approaches that can be used to learn how to manage accreditation properly.

Unfortunately, the demanding level of management and strong leadership of the accreditation strategies, and the fundamental research and writing of the self-evaluation report is not normally performed by a senior college or university administrator because of the large amount of time that is needed on this exercise. Accordingly, key staff and institutional personnel who are very knowledgeable about the institution, and are quite familiar with accreditation and outcomes assessment processes that are important and are to be demonstrated and documented should be designated for this project. The director of Assessment and Institutional Research at the institution in this example, NDSCS, was specifically chosen for this position. The organizational culture was then appraised to find out whether personnel employed there are authorized and expected to perform in a self-directed manner, or whether the managerial decision-making procedure is centralized, with the preponderance of managerial decisions actually performed by only a small number of individuals. Once the solution was achieved, a successful accreditation management strategy that leveraged organizational structure and institutional culture was implemented to address the accreditation standards requirements for evaluation and improvement. The objective at NDSCS was to incorporate the accreditation self-evaluation into the regular operating routine of the campus and to reveal how the procedures used to assess and measure the college were ongoing and renewable once the accreditation process and on-site peer-evaluation were completed.

It is also important to note that support from many campus stakeholders, such as faculty members, administrators, staff, and students, was obtained at NDSCS by utilizing both official and unofficial conversations between those individuals and teams who were creating the accreditation procedures and other important institutional managers and faculty leaders. These numerous discussions helped establish contact with the college campus constituencies about the proposed procedures, generated feedback, and even assisted in discovering strong contenders to work on the accreditation steering committee. The time schedule for this effort was also created keeping the institution's culture and established behaviors in mind, and the management approach for execution was planned accordingly. The stages at NDSCS that led to their victorious accreditation self-evaluation included naming the institution's steering committee, deciding on the information-collection procedure, observing the performance of the steering committee, supplying an accreditation resource room, organizing the writing and editing of the self-evaluation document, and gathering additional reaction on the report from the institutional community. The outcome of these efforts at NDSCS was a triumphant regional accreditation and peer-evaluation team on-campus visit that engaged the better part of the individuals at the institution and was managed with great consideration of the college's culture and capabilities.

There are other important issues to consider aside from properly choosing accreditation steering committee members, such as planning based on (or around) the institutional culture, pursuing stakeholder support, and team building. Many postsecondary institutions may have trouble in managing a unified employee team spirit that is positive and motivating, but some colleges and universities inherently possess employee attitudes that are agreeable to this management goal, and other institutions can benefit from this. The accomplishment of a successful reaffirmation of accreditation at the United States Air Force Academy in May 1999 can be credited, at least in part, to an excellent self-evaluation committee, with participants from varied areas of the institution (Reed and Enger 2000). The experience helped the academy realize many important ideas about accreditation, including some lessons in culture that other institutions also learned, such as the significance of choosing the best participants on the steering committee, scheduling an attainable time frame for the project, encompassing and featuring institutional information, and supplying leadership to the committee on problem solving. Nevertheless, some additional lessons to consider are the importance of maintaining direction without repressing creativity and launching a unified voice in the institution's accreditation review procedure. Because postsecondary institutions are inherently populated and tend to employ personnel who are often highly educated, frequently outspoken, imaginative, and unafraid to promulgate and profess differing beliefs or openly contemplate tangential and expansionary ideas that are (or may be) related to the topics currently under scrutiny, it is in the best interest of the institution and the senior administrators, along with the self-study directors, to guide this brainpower and passion in productive ways to assist the accreditation procedure without restraining the resourcefulness of the college or university community culture. Sufficient time was allowed for the Air Force Academy to initially concur upon a suitable plan and to produce several drafts, which then helped to strengthen the accepted schedule of the numerous accreditation subcommittees under the steering committee. Additionally, the academy explicitly sought to craft the self-study report with a unified voice by assigning it to one or two writers and then requesting that other committee members supply feedback on the comprehensive draft documents that the designated authors created. Although this approach may not be appropriate at all institutions owing to internal politics, it illustrates how the objective to unify viewpoints and writing styles in the accreditation self-study report can then enable realization of the overall institutional goal to obtain accreditation approval and organizational revitalization.

An earlier discussion pointed out how important it is for accreditation management strategies to effectively plan the communication with objectives in mind that benefit the institution and endeavor to modify viewpoints and foster accreditation support. The college and university community can be kept informed by creating an *Information Campaign* (McCallin 1999). For example at the College of Financial Planning, which happens to be a for-profit educational institution, the management strategy included the creation of a campuswide campaign to: "(1) inform the internal constituency about its self-study process; (2) encourage their participation throughout the process; and (3) teach them about requirements of affiliation with the Commission on Institutions of Higher Education of the North Central Association of Colleges and Schools" (p. 302). This college's *Information Campaign* contained three primary sections or features. The first part consisted of publicizing the institution's recently created strategic plan and soliciting feedback from the college community. The second part of the campaign consisted of scheduling open community meetings with presentations conducted by the president of the college or the coordinator of the accreditation self-study to explain the connection of the institution's strategic plan to the association's standards and the current status of the report preparation in the college's self-study process. Finally, the third part of the campaign contained a well-crafted weekly institutional newsletter that was published and circulated over a period of six weeks immediately preceding the on-site peer-review visit to the campus. This campaign newsletter was designed to offer the college community an enjoyable, easy-to-read, and pleasant communication vehicle that addressed many issues, such as fundamentals about the regional accrediting agency, the importance of the process for the college, the functions of the self-study report, composition of the peer-review team, and other related things. The layout and design was attractive and reader-friendly. (A sample of the content of the college's *NCA Talking Points* newsletter is illustrated in Appendix D.) Other colleges and universities examined in this book also used newsletters or websites as important tools for communicating with the institutional community about the accreditation objectives, providing groundwork or education for forthcoming peer-review visits to their campus, and persuading the campus community in a constructive way.

CHAPTER 15

Controlling and Directing Accreditation

For both large universities and small colleges who desire to pursue accreditation from one regional association or add many specialized agencies to their list, managing an accreditation process is a large responsibility and endeavor that can be more successfully controlled and directed by using established managerial techniques. Controlling aspects of this situation include observing and implementing expectations and processes that are required by the accreditation association standards, promoting constructive and discouraging destructive actions within the postsecondary institution, and giving incentives to internal actions that help accomplish accreditation goals and improve the institution. In a research paper that examined the association between regional accreditation and improvement at institutions, accreditation was found to have the largest effect on colleges that were smaller and less competitive and that were examined by the accrediting association more often than once every decade (Robinson-Weening 1995). The research survey also discovered that the schedule for accreditation planning in the institution's ongoing life cycle had a crucial impact on the outcome of the accreditation procedure. These findings should be taken into account while controlling the accreditation management strategies at larger universities, which frequently must go further than the nominal fulfillment of regional association accreditation standards to attain successful public recognition and internal revitalization. Similarly, smaller colleges should contemplate the scheduled time plan of the required accreditation on-campus visits in their institution's life cycle, in addition to planning for additional accreditation reviews instead of the traditional ten-year evaluations.

In order to aid the control of positive employee performance and properly reward productivity, many colleges and universities have created methods of operation that involve thorough planning and forecasting of resource utilization, preparation and development of accreditation committees, and data gathering. Existing institutional research departments can often perform an important function in management institutional self-evaluation procedures and the writing of self-study and required secondary documents (Bers and Lindquist 2002). Many trained experts in institutional research possess useful experience in performing research surveys and customized institutional studies, evaluating survey data from available national databases to supply benchmarking information, validating the reliability of information collected, and producing helpful charts and tables for the accreditation self-study report. In order to manage the most efficient use of the institutional research office, colleges and universities should ensure that the department has the authority to participate in the important managerial decision-making meetings that will take place regarding the accreditation procedure, have permission to use the resources that are required, and most importantly have access to information.

Because of the computer technology and information that are now available, postsecondary institutions have a variety of means available that can provide very useful methods to control and direct accreditation management strategies. Use of information technology in controlling and operating the accreditation process should be promoted, planned, and financially budgeted at the institution. These management control strategies can help guide the accreditation effort's communication with the institution's internal stakeholders through electronic mail (e-mail), accreditation web pages, and electronic learning (e-learning) systems. They can also facilitate contact with the accreditation association and self-evaluation committee leaders and members, and help in directing the accreditation work, drafting and writing accreditation documents, research, information gathering, and data analysis. Consistent and authenticated data have now become the focus in many self-study processes, and peer-reviewers are going to colleges and universities looking not only at the information presented by the institution, but also at what efforts are being performed and what outcomes have been achieved at the colleges and universities after the data have been used (Harrington et al. 2000). The product of this data awareness is that successful management of the collection, distribution, and reporting of information is a critically important aspect in the accreditation procedures at postsecondary institutions. Consequently, it is strongly suggested that this important information and data be made available early in the self-study review process and that the self-study information subcommittee (under the direction of the accreditation steering committee) be prearranged as an initial part of a properly managed accreditation self-study.

The gathering, investigation, and dissemination of internal institutional and comparative data are important for controlling and directing the management of accreditation activities. Ordinarily, this information is required on a variety of topics, such as enrollment data of undergraduate and graduate students, generation of student credit hours by the department, faculty research and service performance, rates of student retention, student graduation ratios, college degree output, the history of professional and support staff positions, classroom utilization, building/facilities planning, traditional library and electronic media use, as well as financial budget reports (Harrington et al. 2000). A large amount of the information that is needed for a self-study endeavor may already be present in some shape at many colleges and universities, often enclosed in common reports such as recurring accreditation association updates; state education department reports; Federal Integrated Postsecondary Education Data System (IPEDS) reports, College and University Planning Association (CUPA) information;

admissions-related institutional fact-books; and cyclic, ad hoc, and other reports. Additionally, because new information technology and services are becoming available and more widely used, a good variety of state and national normative and comparative institutional data sets are accessible on the Internet, together with many websites that offer a variety of viewpoints on the utilization and function of this information. (For example, see the Association for Institutional Research at http://www.airweb.org/links/peers.cfm for a comprehensive list of peer comparisons.) In addition, outcomes and student-learning assessment at the degree- program level has become commonly requested and practically required in accreditation reviews today, and many specialized professional schools or colleges at postsecondary institutions that seek accreditation from specialized associations frequently have the opportunity to access superb models of assessment information, such as AACSB, NCATE, and others. However, as part of an internally developed management accreditation strategy for an accreditation procedure, and for other reasons, an institution may decide to research and create surveys internally. These home-grown surveys normally research and assess institutional levels on the usage and perceived satisfaction with departmental programs and services for gauging institutional unit value. If these surveys are developed internally, it is advisable that the methodology be well planned. Fortunately there is often significant expertise available within the institution to rely on because these academic researchers are quite familiar with effective research survey processes and analysis.

To support information, tabulation, and control of accreditation data, software such as *The Dean's Associate,* which has been purchased by over 300 business schools or colleges that are involved primarily with accreditation by AACSB-International, is available (Octogram 2003). This software program can be utilized by business schools that are pursuing or seeking to maintain specialized business school accreditation, or as an example for future internally developed network programs that can be created at institutions for greater customization. While *The Dean's Associate* is particularly aimed at assisting AACSB-International mission-based accreditation standards, it does offer business schools and colleges some flexibility in directing their self-selected stakeholders. The devised objective of the software is to construct a database management system that is sufficiently accommodating to support many of the management, controlling, scheduling, decision-making, and accreditation-related responsibilities of business school deans. Other examples of useful software to aid accreditation management efforts can be found outside higher education, such as in the health care industry. For example, there is a program offered by a developer of managed- care software and a popular physician search engine called GeoAccess (PRNEWSWIRE 2000). GeoAccess established a partnership with CSI Advantage, producer of an integrated, all-inclusive tool for accreditation by the National Committee for Quality Assurance (NCQA). The GeoAccess product provides CSI Advantage's Accreditation Information Management System (AIMS) to managed-care organizations (MCOs) in the health care industry that are seeking to reduce costs by cutting the time and effort currently needed to prepare for NCQA accreditation by half. The offering of this software program may signify that there is a possibly strong market base of customers for accreditation software in the field of higher education. In a market-driven economy such as postsecondary education, this need will probably be filled shortly by a variety of competitive software products. However, for the time being, directing and controlling the management of accreditation processes in higher education is reliant upon the innovative use and practical improvement of existing services and management information systems at institutions.

An informative example of this approach using this type of information system occurred at the University of Cincinnati, where a pattern was used to gather short narrative statements from different academic and support departments for their accreditation self-evaluation in 1999 (LeMaster and Johnson 2001). The university chose to implement a special emphasis or focused accreditation option with the objective of using the reaffirmation process as a means to assess the value of continuing institutional priorities in the areas of teaching and faculty research. The University of Cincinnati is a relatively large and complex postsecondary institution with a variety of missions among its 16 colleges. There is a great deal of autonomy and self-direction in the many college subcultures, which produce many internal institutional obstacles to centrally driven initiatives such as the strategic and operational management of accreditation processes. The planning of the self-study mandated that the university determine a procedure to manage and encourage active participation from all of the academic and administrative support departments in a controlled and scholarly manner. The method that the institution used was an appeal to all academic and administrative support departments to document and put in writing a sequence of brief descriptive statements that were based on models supplied to them on computer format. This straightforward yet successful method included detailed instructions to the various department leaders directing them to contribute by way of supporting information and documents, and that they should respond only to the parts that were pertinent to their area. Each of the issues for teaching techniques and faculty research activities (such as scholarly work and creative activities) included an appeal for instances of current initiatives, results that validate the effect of the practices on student learning, and reveal how the illustrations which are shown have been integrated within the department and institution to improve performance. Additionally, a summation analysis was also gathered on the strong points, areas in need of improvement, rising issues, and intended new approaches for teaching techniques and faculty research in the many departments and administrative areas. This information could then be proactively utilized by the accreditation self-evaluation committee in the overall strategy for appropriate arrangement of public recognition, as well as internal institutional renewal plans throughout the university.

Comprehensive university or collegewide accreditation management strategies and plans that are developed by the academic leadership are essential not only for supplying guidance, but also for controlling and directing the operational approaches, which can be decentralized in many instances for maximum effectiveness. Postsecondary institutions have the opportunity to emphasize information-gathering systems such as the type illustrated previously. Supplementary purposes can be included by connecting goals with accomplishment assessment and accreditation standards. One illustration of a creative method that successfully connects an institution's strategic goals such as important accreditation achievements or renewals to detailed performance results is seen at Regent University in Virginia (Maher 2000). A well-planned, long-term, adaptable, annually renewed, five-year strategic planning procedure is connected to the university's institutional mission and vision, yet is controlled using processes that link detailed outcome events to financial distributions within the institution. The model for this method to control responsibility, and the ongoing development contains four elements: arranging, gathering information about the activities, evaluating and measuring the completed actions, and official suggestions for improving subsequent activities on each of the schedule procedures.

In order to control and direct the efforts of effective accreditation management, the various academic departments and support units at Regent University had their information system worksheets files saved on the campuswide institution network. Located within each of the spreadsheets for the strategic planning process are data columns (also called computer "fields") that necessitate detailed information to follow the four components mentioned earlier. These data fields in the strategic planning worksheet contain a range of items that lend material and precision to the planning structure for greater control of the procedure, with field labels such as "objective number" (which refers to specific numbers in the institution's mission such as 1.01), "objective statements" (which refers to one of the 22 stated goals), procedures to be initiated, employees who are responsible, the planned date for completion, anticipated outcomes, net revenue surplus or cost, and priority (such as "Core," "A," "B," or "C") as established by the official budgeting guiding principles (Maher 2000). In addition, there are also data columns or computer fields for the standards of the regional accreditation association (in this case, *SACS Criteria for Accreditation, 2002*), and this compels linking the particular action being referenced from the institution's goals directly to the accreditation standards to which they pertain. After the information is gathered on the computer worksheet from the various departments, the overall plan is then connected to the financial planning process in a highly controlled methodology. The anticipated outcomes concentrate on specific performance results, which are not just the procedures or processes. For instance, "Eighty-five percent of School of Law students will pass the ABA exam" (Maher 2000, p. 31) is an illustration of a programmed result.

Controlling has been defined as a procedure for assessing accomplishment and then implementing decisions to assure desired results (Schermerhorn 2002). For properly managed accreditation strategies, this involves using representative yet critically imperative processes to stay in communication with internal and external community members of the institution, keeping updated regarding operations being performed, utilizing management information systems efficiently, and maintaining the accreditation objectives that include public acknowledgment and internal institutional revitalization frequently on the agenda. Nonetheless, these control and direction fundamentals of proper accreditation management are still just one section of the comprehensive strategy. Indicating and presenting the institution's performance to the accreditation association and others, in addition to sufficient financial allocation, are also very influential issues that must be understood.

CHAPTER 16

Writing the Self-Study Report

Accreditation managers will find that nearly all accrediting associations provide instructive handbooks and assistance for postsecondary institutions that are seeking accreditation to support the self-study preparation and the crafting of required reports and other necessary materials. In examining the objectives of various accreditation associations, they typically seek to connect the self-study to the standards of the agency, internal expectations, and external requirements. Regardless of whether the college or university is conducting a traditional comprehensive-type, special-emphasis, or collaborative self-evaluation study, the self-study report should be managed properly by implanting the managerial elements that have been discussed earlier. Lately, the accreditation association criteria for accreditation of many associations has strongly emphasized the utilization of student- learning assessment and development as a primary measure of institutional condition and value, along with faculty performance and related features of the learning environment, such as the academic curricula, student extracurricular activities, and campus facilities that promote the physical, community, spiritual, and intellectual advancement of the students (Braskamp and Braskamp 1997). Several researchers, such as Barr and Tagg (1995) and others, have recognized this development as a recent update from viewing the postsecondary institution as not merely a place of instruction, but also as a center for producing learning (Barr and Tagg 1995). Some accreditation management strategies that colleges and universities can leverage to meet and exceed these expectations are: judiciously selecting the terminology and phrasing employed in the self-study document, being adamant on expectations without homogeny, utilizing a wide selection of evidence, concentrating on academic excellence, and measuring the entire range of student learning and personal development (Braskamp and Braskamp 1997). The institutions under review can also be innovative in distinguishing their quality based on student achievement over time, often by maintaining the perspective of the overarching self-study objective to communicate institutional value to the outside evaluators and utilizing representations or examples of excellence. This does not mean that an institution should be untruthful in the accreditation report, but it is wise to thoughtfully choose the words that are used and performance that is documented and publicized in ways that help achieve the external approval sought by the accreditation association while revitalizing the college's or university's internal assurance to the renewed, or even initially formed, student-learning goals and assessments. Individual faculty members should be included in evaluating the quality assessed, not only to obtain their support for this rejuvenation of institutional values that were accounted for and pursued, but also to share the academic character of the institution being examined with the outside peer-evaluators. The decision of academic worth and excellence has considerably more influence than just administrative edicts from above, or other strategies that do not put together significant accumulation of education and knowledge within most colleges and universities. Consequently, strong attempts should be made to help guarantee that faculty members comprise a large segment of the institution's self-study accreditation steering committee and various subcommittees.

Probably the most important factor in successfully navigating an accreditation process is a written self-study report that is required by the accreditation association. This report is a major means of transmitting the information about the accreditation strategy that is being put into practice. It is a large project to properly plan the documentation of accreditation management strategies that address accreditation standards. It involves preparing a book-length document on the internal accreditation steering committees' conclusions concerning the institution's strengths and weaknesses and making suggestions within a fairly tight schedule. Initial conclusions in the planning of the self-study report management plan are crucially important since they administer the complete endeavor of the self-evaluation during its standard 18-month to potentially 36- month (three-year) length of time (Reich 1999). Obtaining learned counsel and wide response in the first part of the accreditation planning and initial drafting of the procedure is enormously beneficial because the importance of this advice multiplies, in that it enables successful editing, rewriting, and further planning. It was stated by a regional accrediting association that

> the self-study report is the document that summarizes each institution's self-analysis and future plans. It sets the agenda for the visiting team of peer reviewers. More importantly, it sets the agenda for the institution itself for several years. As a "living" document, a clear self-study report should serve as a plan and a reference source for all of the institution's constituencies. (MSACHE 2002, p. 59)

The written and electronic communication efforts are a significant part of properly managing the accreditation effort to effectively inform the various stakeholders involved in the process. In crafting the accreditation report, institutions should understand that content is not the only important aspect; how the content is presented is also very important. As previously mentioned, those colleges and universities that authorize conceptual efforts and communication techniques that direct and translate various organizational changes for members are probably going to be more durable when confronted with outside trials (Chaffee 1984).

There are many functional and applied issues to understand while appropriately crafting the self-evaluation report, interim updates, accreditation follow-up communication, replies to peer-review teams, and other interactions with accreditation associations. One author of a successful accreditation self-study report who also participated on various peer-review visit committees as a consultant-evaluator for a regional accrediting agency, and who is a professor of English and department chair at an institution, states that the tone of the writing, grammatical choice, technicalities, punctuation, and presentation are basic yet crucial issues that should be thoughtfully planned and implemented (Swanson 1999). Swanson proposes that the writer's tone should be truthful, lucid, and succinct, and that the self-evaluation report and other written exchanges should contain no misinformation and useless jargon. This is sound advice, and accreditation leaders and individuals participating in accreditation efforts can adhere to these procedures while writing and leveraging an effective accreditation management strategy that strives to influence institutional perspectives and outside opinion, enhance integrity, and use management techniques in thoughtful and honorable ways. This approach may include emphasizing college or university strong points in several areas, tackling weaknesses forthrightly with detailed plans for advancement, and crafting the working in a style that is predictable or unpredictable from an institution-type of perspective to peer-evaluators in another perspective. Even though this advice may not appear innovative to accreditation managers at some institutions, Swanson's suggestion to be conscientious in utilizing proper grammar, writing mechanics, and mundane details such as punctuation are truly important to follow-through with. This is especially important in frequent problem areas such as fragmented sentences, exceedingly long sentences, proper tense, person uniformity, abbreviation, agreement, and modifiers. Granted that while some readers are often more critical than other less detail-conscious readers, minor details such as these can either emphasize the institution's intended accreditation strategy or cause such a reader to become distracted by problems in the format of the writing.

At the beginning of an accreditation self-evaluation procedure, there are important issues to consider and plan for and external examples to learn from. Often, individuals charged with writing a chapter or overseeing the entire process may believe that there is a true "right" and "wrong" in crafting this important document, but this is not true. An experienced accreditation consultant-evaluator recently wrote that there is no one truly correct way, or even finest method, of writing the report, or actually anything at all to do with the self-study process (Carroll 2003). This is a good point, and it is also exemplified later in this book with the illustration of new approaches to accreditation using business quality initiatives and new information system technology. The experienced reviewer mentioned (Carroll) and others (Noonan and Swanson 2003) have offered specific suggestions (listed below) to review committees and subcommittees charged with crafting the self-study report. Overall they advise other institutions to find their own best way to meet the stated accreditation criteria, but carefully write the report by keeping the reader as well as the institution's goal in mind. Here are some useful suggestions for institutions preparing for report writing to consider:

- Ease of Comprehension. Since the report will be read by several types of readers, including internal institutional leaders and external reviewers, each topic should be addressed in a way that a reader who is not familiar with the institution can comprehend the content. It is likely that there will some content that needs to be clarified or elaborated on when the peer-review team visits the campus, however if too much explanation is needed then the team will waste unnecessary time on information clarification and not enough time on in-person interviews to gain a full appreciation of the college or university. One seemingly simple element of advice to remember is to limit the use of abbreviations that may be very familiar to local campus individuals but are unknown to outsiders. This is quite common for locations, committees, institutional processes, and other portions of an institution that community members may use frequently and has become part of the lexicon.
- Comprehensiveness. The document should include all new and significant developments at the institution, both the positive items as well as the negative. The external peer-review team would find it very frustrating to arrive at the institution for the evaluation visit and discover that the self-study was not a complete review and only completed just a few weeks before the visit. The self-study should illustrate progress that has been made to address the weaknesses at the institution, and those mentioned in prior accreditation reports. These portions of the report do not need to be lengthy, but they should be unambiguous and direct in providing evidence.
- Minimal Repetition. Even though it is important for the self-study to be comprehensive, it is also necessary to avoid repetition. A certain amount of repetition is unavoidable in any large document such as this, since the nature of the accreditation criteria and questions about previous evaluation visits may require including the same points more than once in the length of the text. One institution that underwent a review included its mission statement several times in response to various questions, occasionally providing some amplification. When the consultant-evaluators representing the accreditation agency finished reading the self-study report, they knew the institution's mission quite well and better than their own. The institution was found to have a clear sense of the mission, but the writers of the self-study seemed to be unable to find alternative ways to answer some of the questions.
- Careful Editing. As the drafts are written, one or two individuals should be appointed as editors to review the content and style and look not just for grammar errors but also for repetition, errors of content, and style and so on.
- Grammar, Mechanics, and Punctuation. As mentioned earlier, the document should be carefully scrutinized for errors such as sentence fragments, run-on sentences (or comma splices), inconsistency in verb tenses, shifts in person, lack of subject-verb agreement and antecedent-pronoun agreement, confusing and misplaced modifiers, incorrect punctuation, and confusing abbreviations.
- Writing with One Voice and with Coherence. The report should read as if it is written by one person, and not look like a piecemeal document that has been

spliced together from several sources. Then the report is near completion, someone who has not been intricately involved in the accreditation report writing should be asked to read and review the document to look for coherence and determine whether it speaks with one voice or a cacophony of different voices. If it does, then it should be fine-tuned and edited as needed so that it appears to be cleanly written.

- Tone. The tone of the self-study narrative indicates the attitude of the writer toward his or her audience. Writers should strive for a tone that is honest, free from propaganda, is clear and concise. In order to achieve such a tone, writers must remember that their readers are peers, fellow educators who will understand some the necessary jargon of the profession, but may not know an institution's particular jargon and abbreviations. Therefore the writer must walk a fine line to avoid a patronizing, lofty tone, and at the other extreme, a too-familiar, chatty one.
- Clarity. Writers should avoid wordiness, and use active rather than passive voice. When subjects are actually present in sentences and are doing the action, the work of the sentence, the document is more interesting to readers and lively than when objects receive the action. Writers should also check for parallel construction, by making sure that all similar parts of a sentence, a list, a table, a chart, or a graph are in the same form. Attention to form will minimize confusion of the readers. In addition, there should be variety in both syntax and diction.
- Presentation. A table of contents should be provided for the readers, to make the report as clear and readable as possible. Tables of contents that cover several pages and are cumbersomely numbered should be avoided. Section headings (with page numbers) at the tops of pages are helpful. In addition, a list of appendices, materials to be found in the resource room (including the e-learning system used in the accreditation process and/or virtual resource center that will described later in this book), and other supplemental material are helpful to the evaluation team reading the report. Also, margins should be consistent and the same format should be used for all headings throughout the document. (Adapted from Noonan and Swanson 1993, used with permission)

It is also advisable to provide a style sheet for those who are writing and for the editor/chair of the steering committee to use (Carroll 2003). Occasionally, steering committee members are directly assigned to obtain the relevant information from institutional departments and then craft the report chapters. In other instances, the various academic departments, committees, and others actually write the report and then one individual, such as the self-study steering committee chair, performs the final writing and editing. The chair should actively look for challenges to the institution as well as its strengths and include them in the self-study report because this is not completely meant to be a public-relations document (even though the achievement of accreditation is useful for achieving public recognition). Editors should look for inconsistencies in the data, which often come from multiple sources, and this can be addressed with the information technology described in Part Four of this book. Although acronyms should be avoided along with other jargon, if they are used at all they should spelled out at the first occurrence of each chapter. It can be very confusing for external peer-review evaluators to see many acronyms and not understand their meaning after reading so many reports. In addition, other advice from experienced voices in this field include making several back-up copies of the report, in addition to the centralized information services that back up the network version, because the network professionals cannot completely guarantee the back-up but accreditation chairs can do this easily with their own computers. The chairs should also update the data on the self-study just prior to the visit, especially if the report was written quite a while before the external evaluation team's actual on-campus site visit. Individual chapter drafts can be circulated to various internal institutional departments as they are written; however, it is better to put the final draft online for overall internal campus use and critique. However, this should include paper copies as well as online versions for those individuals who may not be computer literate or have access to the computer versions. Final advice on writing strategies include providing an executive summary to institutional board members (because they are unlikely to read the entire document) and conducting thorough communication during and at the end of the self-evaluation process to resolve any outstanding issues that were raised during the self-study process (Carroll 2003).

There is also specific advice to institutions that are writing a self-study report for a focused visit (Noonan and Swanson 2003). In some situations the self-study approach may not be appropriate for a focused visit as it may be too narrow or too broad in strategy. The self-study for focused reviews should not merely be a shorter version of the comprehensive report, or merely a brief discussion of a narrow interpretation of the two or three specific areas of the focused review. Some focused visits are conducted to explore a new degree or change, such as the addition of a new master's degree or a new capital campaign. There should be sufficient discussion of the institution's overall background in order to demonstrate that the addition of the new degree program or capital campaign will be compatible with the institution's mission and abilities, and to determine whether the new endeavors will strain the human, financial, and organizational resources of the college or university. In some focused reviews, multiple new degrees or other issues are examined, and it is important for all programs to have the proper background research, supplemental research documentation, and realistic basis for achievement of the proposed improvements to the institution.

In summary, when writing the accreditation report, the documents that institutions prepare for submission should be written very clearly, organized properly, concentrated on critical topics that were thoroughly deliberated, and crafted with the knowledge that it will be used to inform a diverse assortment of internal and outside groups (Kells 1994). In this era of change in higher education, those outside academe are gradually expecting to find more information and documents such as college and university self-evaluation reports on the Internet for the public at large—prospective financial donors, state offices, as well as other accreditation associations for whom the self-study report was not designed—to examine. As a result, the substance, as well as the form, of the written presentation is important, along with an understanding that the accreditation association or agencies for which the report is prepared may not be the only recipient or reader. Additional preferred qualities of the self-evaluation report include ease of reading, succinctness (often 100 back-to-back single-spaced pages is an entirely suitable length needed), and an evenhanded perspective about the institution or program being evaluated, written from the perspective of a unified voice. Of course, it is vitally imperative to incorporate a thoughtful explanation in regard to how the association's accreditation standards are more than satisfied by the institution, together with an arrangement of the college or university's appearance that is suitable to the accreditation management strategy of positive and encouraging external viewpoints.

CHAPTER 17

Budgeting for Accreditation Management

In regard to budgeting of accreditation processes, one of the oft-cited criticisms of accreditation by those in academe is the relatively large expenses that institutions must set aside for their required regional and increasingly needed specialized accreditations (Graham et al. 1995; Trow 1996). Although new and innovative accreditation strategies can be utilized to help reduce these expenses, such as those discussed in the planning section of this book on accreditation synchronization, the inclusive objective of budgeting for accreditation management should contain detailed and comprehensive plans for complete budget expectations to attain the accreditations that are needed in the institution's overall strategic plan. Additionally, the increase in expected tuition and other income for the institution that can be achieved for the internal revitalization and external public recognition that accreditation can accomplish should also be concurrently planned for in the budget. For the majority of postsecondary institutions, the expense for regional accreditation is a truly required expense to maintain student enrollment due to the current connection between regional accreditation and approval of federal student aid and grant funds. However, specialized or professional accreditation can impact institutional budgets in several ways. It is thought by some in higher education that professional accreditations may restrict an institution's planned faculty utilization and affect student admissions, which can result in increased expenses and reduced revenues. However, other specialized accreditation achievements may produce greater public approval and awareness in certain fields, which then is the source of more enrolled and paying students for the budget. A number of professional accreditations are actually required in some states—for example, NCATE specialized accreditation for teacher programs. Therefore, a variety of budgetary management issues need to be considered when designing accreditation plans, along with creating short-term and long-term accreditation goals.

Effective accreditation management techniques to consider in budgeting accreditation efforts include not just particular self-evaluation expenses, but other payments such as those connected with the outside peer-review committee visit, internal and external contacts, communication, travel funding, site visit particulars, and preparation of the institutional community (Palmer 2002). The various stakeholder groups at the institution need to be properly informed regarding the pending external peer-evaluation team visit to the campus, scheduled dates, visitation team composition, the expectations of the team and the institution, topics that could pose challenges, and broad knowledge regarding the objectives of the team visit and the overarching goals of the accreditation management strategy. For instance, at a relatively large, urban, multicampus institution that underwent a comprehensive site visit in 1999 by a regional accreditation agency, the self-evaluation committee leaders guaranteed that the institution's faculty members, administrators, and staff were properly informed about the accreditation site visit by making presentations, distributing important information to academic program advisory committees, different divisions and academic departments, and groups within the community. To help ensure that peer-review visitation team members were properly supported in their responsibilities in the evaluation and to show that the institution being examined was fully accommodating the team's needs, a dedicated meeting room with appropriate information technology was arranged both at the hotel and on-campus. Other planned budget expenses that were related to the on-campus visitation were a team hospitality room at the institution and preliminary planning presentations for the institutional stakeholders. Successful budgeting for managing an accreditation process entails comprehensive planning and even creating elements to guarantee achievement of accreditation, success of external accreditation agency approval, and positively influencing the viewpoints of the internal and external participants, as planned in the overall accreditation management strategy. The result of the peer-review team visit to the institution illustration described here was a positive achievement, and due largely to proper financial expense planning for the internal communication aspects and external team on-campus arrangements.

This part of the book on strategies for achieving accreditation articulated a schema for using accreditation management strategies derived from established management and higher education literature, and described accreditation best practices that were obtained from journal articles and conference proceedings. The outcome of actually implementing, or perhaps not using at this time, the established and innovative management approaches that can be adapted from the field of business is not as important understanding their possibilities. If these particular management tools become an integral part of many institutions' strategies, there is no denying that they will change the way fundamental concepts of business and public management are being implemented at educational organizations to meet survival and growth requirements, and, most importantly, to address the great challenges that are confronting higher education today. There are additional concepts that can help postsecondary institutions to obtain public recognition and internal revitalization through academic accreditation endeavors, such as relatively new business quality approaches and dynamic institutional leadership.

PART FOUR

New Strategies for Accreditation

CHAPTER 18

Using Management Quality Techniques for Accreditation

An important characteristic of the latest generation of postsecondary accreditation is the application of business quality initiatives, for instance, the Academic Quality Improvement Project (AQIP) and the education category of the Malcolm Baldrige National Quality Award (MBNQA). These approaches were first developed in the field of business management; they have been used by many institutions of higher education in recent years in certain areas. Some of the US regional and specialized accrediting agencies now require institutions to use these approaches, partly to address the public calls for greater levels of accountability of colleges and universities today. The higher education scholar, Martin Trow (1996) states that provided academic accreditation is viewed as the process by which higher education qualifies and polices itself as opposed to the alternatives, superior systems of improving academe may not receive full attention. Nevertheless, enhancements and other options for the required accreditation self-reviews are currently being implemented by a large number of accreditation associations and institutions being reviewed, and will be investigated in this chapter. Trow also suggests that a more analytical self-study of existing internal institutional quality management processes be considered, perhaps looking at the examples from the international higher education community.

In the U.S. system of higher education and elsewhere, the accreditation system has started to implement new business quality strategies as an option to enhance the traditional system. These approaches have elements similar to academic audits in that thorough self-evaluation is performed, yet an exhibition of increased productivity and excellence is also needed. The relatively recent business quality strategies are being evaluated and used to some extent by postsecondary institution leaders around the United States and globally as higher education is progressively more inundated with external expectations to display competence and overall value (Beanland 2001; Janosik et al. 2001). For instance, the government of Australia has established the

Australian Universities Quality Agency to encourage quality results within the postsecondary education institutions that they hope will manage overall educational organization enhancement and expansion (Beanland 2001). Across the ocean, the institutions in the United States have employed a number of quality improvement strategies, such as the Academic Quality Improvement Project and the Malcolm Baldrige National Quality Award Education Category (NCA 2001; NIST 2001). The AQIP is designed to provide an "innovative, more challenging alternative process for reaccreditation" (NCA 2001, p. 10). The Baldrige Award determines and sets forth standards that are comparable to some extent with accreditation criteria in academe, and are acknowledged by various state and national education award programs.

The University of Wisconsin-Stout (UW-Stout) became the first postsecondary institution to win the Baldrige Award in 2001. As mentioned earlier in Chapter 3, the Baldrige Award criteria have a total of seven categories: leadership; strategic planning; student, stakeholder, and market focus; information and analysis; faculty and staff focus; process management; and organizational performance results (Sorenson et al. 2002). As the basis for fundamental self-evaluation and development, UW-Stout began using the Baldrige Education Criteria for Performance Excellence in 1999. An examination of the methods used by UW-Stout to meet the Baldrige criteria reveals that this accomplishment was largely the result of careful and positive communication of current methods at the university and supplying lucid and coherent assertions and assessments. An example of this approach can be seen in the organizational leadership at UW-Stout. It consists of a set of standard, hierarchical higher education functional areas led by a chancellor who is charged with achieving the institution's objectives established by the board of regents. The approach, not uncommon in colleges and universities, follows a mechanism of shared governance that includes a faculty senate, senate of academic staff, and a student association. This arrangement is designed to promote responsive interactive communication and basically flattens the university's organizational hierarchy by implementing broad participation in all areas of institutional governance. In regard to the criteria for strategic planning in the Baldrige standards, UW-Stout stated that their strategic goals already include factors such as excellence in faculty teaching, scholarship, research, and community service. An example demonstrated how, through the institution's use of feeder schools, employers, alumni, the UW-system and their board of regents, student, stakeholder, and market focus are key elements of success. Additional Baldrige Award criteria, such as faculty and staff focus, information and analysis, process management, and organizational performance results were likewise addressed by openly reporting current institutional methods and outcomes that are sometimes quite common in other postsecondary institutions. These common organizational procedures, interactions, and configurations demonstrate that if a recognized institution such as UW-Stout can succeed in meeting the Baldrige Award criteria, then other established colleges and universities can also strongly consider applying for and eventually achieving this supplementary and impressive external recognition, or at the very least obtain the gains from the application process.

The Baldrige Award process and regional accreditation such as the NCA are performed as groundwork for an on-campus evaluation visit by a committee of peer-reviewers chosen to assess the institution and recommend its suitability to be an affiliated member of the accreditation association, and, in the process, for the Baldrige

Award, (Brewer et al. 1999). As part of a comprehensive approach using total quality management, the Baldrige Award is perceived by some individuals and leaders as the highest recognition that an organization in the United States can attain. However, the Baldrige Award program also has some critics. According to them, the procedure is expensive because it establishes a competitive environment and, in reality, an applicant's desire to win is an important reason for achieving the award (George 1992). The skeptics also see problems in the way many organizations that engage in Baldrige Award programs appoint external consultants to pilot the procedure. However, despite these problems, many postsecondary leaders and educational institutions believe there is great value in the procedure and are seeking this important outside recognition especially if it is coordinated with accreditation reviews.

An example of an institution that successfully leveraged the Baldrige criteria for a regional accreditation self-study is the Mt. San Antonio College. It is a large southern California community college district with an enrollment of 22,500 FTE (Full-Time Equivalent) students (Feddersen 1999). When the first National Baldrige Education Pilot Criteria for Education and Health Care was published in 1997–1998, the Mt. San Antonio College was planning to perform a regular comprehensive self-study for renewal of regional accreditation. Fortunately, the regional accreditation agency (the Western Association) had begun to promote and support special self-evaluations for institutions that were being reviewed. Mt. San Antonio submitted an application and was granted authorization to utilize the Baldrige Award criteria as the outline for their self-evaluation. The college then integrated the standard accreditation requirements with the Baldrige Quality criteria. Incidentally, Mt. San Antonio was only the second Western Association community college to conduct an alternative self-study and the first to use the Baldrige criteria as the basis for evaluation. The college structured the self-study around seven special study committees, consistent with the appropriate Baldrige criteria, similar to the way conventional comprehensive self-studies for many regional associations often create teams based on the stated accreditation standards.

Following the application of the management quality strategies, Mt. San Antonio College recognized important systems and essential processes that supply the basic structure for the successful performance of the institution. The Baldrige program procedure tends to concentrate on how customer and important related processes are defined, how significant customer necessities are determined, how key assessment gauges and objectives are decided, and how functioning is calculated (Feddersen 1999). The president of the college affirmed that the preparation, direction, and analysis the institution obtained by participating in the Baldrige program and from the external consultant was tremendously beneficial during the self-evaluation procedure. However, it should be noted that not every educational institution that uses a Baldrige strategy in higher education utilizes a consultant. The president of Mt. San Antonio College stated that combining the accreditation standards with the Baldrige criteria was challenging. He listed a number of important lessons that the institution learned from this innovative and complicated procedure that may not be easy to understand at first, and can be difficult and frustrating as institutions imbibe by implementing.

- **We learned the value of applying a systems perspective.** Accreditation standards are independent variables. Each one stands on its own. The Baldrige asks a set of common and related questions that form a thread that is woven through all criteria. It forces connections and integration and emphasizes how everything contributes to results. By probing deeply, asking many "how" questions, we were led to ask "why." By forcing us to explain things, certain truths were revealed.

- **We learned the critical role played by Key Performance Indicators (KPIs).** We had not clearly identified KPIs, and didn't fully understand them or how they linked operations to results and improvements. Now we can see how performance indicators are related to trends, benchmarking, planning, and goal setting.
- **We learned that measurement and benchmarking are the drivers of change and continuous improvement.** Our Baldrige Self-Study pinpointed our lack of metrics and outcome information. We also learned that we were not very sophisticated about benchmarks and benchmarking.
- **We learned the importance of leadership system.** While this may seem like an obvious finding, in many institutions leadership is narrowly defined and often misunderstood. The Baldrige framework taught us to look at shared leadership as a system and how it connects to and drives other parts of a larger system.
- **We learned the importance of training.** Using the Baldrige criteria to assess where an institution is and to drive performance improvement will not happen unless most staff really understand systems thinking. In conducting the Baldrige Self-Study we had to learn how to answer the many similar, but slightly different, "how" questions. We also had to learn how to ask the right questions if we were going to collect useful data.
- **We learned that the Baldrige seems to generate more gaps—more areas for fundamental improvement.** It's not just that the Baldrige raised a lot more questions with fewer items than accreditation; those probing questions appear to generate more fundamental concerns. They really showed us what was missing. For example, we looked at how we do planning and then linked that to results. Baldrige appears to probe deeper, to expose more of the disconnects in the institution. For instance, while accreditation asks whether a certain policy is in place, Baldrige asks how the policy was devised and how the institution evaluates the effectiveness of the policy on student learning. (Feddersen 1999, pp. 14–15)

Even though Mt. San Antonio and the institution's chief executive officer reported that integrating traditional accreditation standards with Baldrige categories created a substantial quantity of additional work, but did not automatically increase the usefulness of the self-evaluation, they did learn that it made the process more important and was overall a valuable endeavor. Generally, the college found that employing the Baldrige Award program was a superior institutional evaluation and development strategy. This method can be considered an effective accreditation management technique that influences the opinions of important community members and the practices in a variety of areas that are central to the attainment of regional accreditation and revitalization of the institution. Obviously, merely establishing the Baldrige criteria within a postsecondary educational institution such as a college or university does not automatically mean that accreditation approval will be obtained. One obstacle or hurdle in utilizing the Baldrige Award criteria is recognizing the talent and expertise needed for the institution's employees to successfully perform their roles and factor in meeting the criteria. It is consequently imperative to identify the principal community groups within a college or university as connected to the stated criteria, and follow up with a list of talents or capabilities required by these groups. Careful planning is highly desirable. A case in point can be observed at Northwest Technical College, Minnesota. Their renewal of the regional accreditation self-study procedure integrated the Baldrige criteria and focused on skills that are important or crucial to successful instruction and student learning (Swanson et al. 2000). Table 5 illustrates the institutional community groups as defined through collaborative endeavors of the college's senior administration cabinet, faculty senate governance structure, student senate organization, and the general advisory committee. In addition, the table specifies the areas of primary convergence for each important stakeholder or community group.

Every stakeholder group at the college was invited to supply a list of suggestions for enhancing each of the performance indicators that were recognized in an assessment. Once the set of evaluation criteria was established and assessed, performance indicators that were considered to be very important student competencies were incorporated into the assessment together with criteria for the Baldrige Award program. This was thus arranged to set up guiding principles for employee development policies. The institution discovered that contributing faculty members and other employees found the procedure not to be intimidating and productive for the institution.

In contrasting the Baldrige Award education criteria and the standards set by a representative regional accreditation association, it can be clearly seen that there are a lot of similarities. For instance, both sets of standards are technically not optional for the institution; possess unambiguous principles or criteria; entail written self-evaluation reports and comparable objectives; and result in a relatively high status connected with their attainment (Bishop et al. 2000). These review methods possess several status levels, and they individually require the applying institutions to gather information elements. Additionally, both require that the regional accreditation procedures as well as the Baldrige program have ongoing quality improvement as a primary objective. Both methods include on-campus visits by selected reviewers.

Table 5 Stakeholder Groups for Baldrige Criteria

Primary Focus Matrix Baldrige Performance Criteria	Administration/ Cabinet	Faculty Senate	Student Senate	Academic Deans	Faculty	Academic Affairs Support Staff	Student Services Support Staff	Maintenance Support Staff	General Advisory Committee
1. Leadership	X	X		X					
2. Strategic Planning	X	X							X
3. Student and Stakeholder Focus			X		X	X	X	X	
4. Information and Analysis	X	X	X						
5. Faculty and Staff Focus	X	X	X	X	X	X	X	X	
6. Educational and Support Process Management			X	X	X	X	X	X	
7. School Performance Results	X			X	X	X	X		

Source: Swanson et al. "Incorporating the Malcolm Baldrige National Quality Criteria into Your NCA Accreditation Process," p. 49 (used with permission)

There are also a number of differences between these programs, such as the unique point value system that the Baldrige Award program utilizes, and nearly no currently established accreditation reviews require. The traditional system of regional postsecondary accreditation in the United States has a respected and lengthy history that goes back well over a century in time, while the Baldrige Award program is a relatively new approach to quality assurance. There are further differences in the two systems such as the annual review period that is required for the Baldrige Award, while, typically, traditional academic accreditation requires a five- or ten-year interim between evaluations. Moreover, postsecondary accreditation procedures were developed particularly for higher education while Baldrige has its roots in the field of business management and quality improvement.

The majority of principal distinctions between academic accreditation standards and the Baldrige Award criteria centers on the institutional factors being assessed and who at the institution guides the advancement (Bishop et al. 2000). While a regional accreditation association such as the North Central Association (NCA) appraises current performance, and the enhancement is at least partially guided by faculty members and students at the postsecondary institution, a Baldrige Award evaluation determines how this is accomplished by examining the tactics, operation, and outcomes. For instance, table 6 illustrates a representation of the primary and lesser associations between the Baldrige Award criteria and the NCA accreditation standards.

The first Baldrige criterion is leadership. It fundamentally compares with the first NCA criterion that is focused on an organization's mission. The second Baldrige criterion, strategic planning, is likewise associated with the NCA's fourth criterion that concentrates on planning. The other criteria are similarly matched for the two organizational evaluation systems in a very effective manner that other institutions can use as a model to link the systems.

Table 6 Baldrige Criteria Compared to NCA Criteria

Baldrige Criteria	NCA Criteria				
	Criterion 1 Mission	Criterion 2 Systems	Criterion 3 Assessment	Criterion 4 Planning	Criterion 5 Integrity
Criterion I Leadership	•	•	•	•	•
Criterion II Strategic Planning	•	•	•	•	
Criterion III Student and Stakeholder Focus	•		•	•	•
Criterion IV Information and Analysis	•	•	•	•	
Criterion V Faculty and Staff Focus		•	•	•	•
Criterion VI Educational and Support Processes	•	•	•	•	•
Criterion VII School Performance Results	•	•	•	•	•

Source: Bishop et al., "Do We Get Them?" p. 42 (used with permission).

An example of a postsecondary institution that leveraged the innovative strategy of linking their accreditation self-study review for the NCA regional association with the Baldrige Award program to establish a comprehensive strategy of integrated ongoing improvement was illustrated at the Wisconsin Technical College System (WTCS) in 1999 (Bishop et al. 2000). This background actually began to emerge in the early 1990s when the WTCS appointed external consultants from the Community College Consortium of the University of Michigan to create an individualized institutional effectiveness model for the WTCS. This model was founded on the then currently used assessment activities and was intended to be altered by the individual institutions involved.

The institutional effective model uses core indicators that can assess and evaluate student accomplishment and institutional performance, in addition to responding to public expectations regarding accountability, the NCA accreditation standards, as well as the documentation reporting required at the federal and state levels. A total of 17 core indicators of institutional effectiveness were identified. These included ascertaining student needs; detecting student functional skills on entry to the institution; measurement of course completion; assessing student grades; gauging student satisfaction; evaluation of student retention and withdrawal statistics; graduation rates; student knowledge at exit; and public satisfaction (Brewer et al. 1999). These indicators were selected through a procedure involving a variety of groups and information sources to discover measures that effectively describe the institution's performance. Because these core markers supply a gauge to measure institutional effectiveness and accountability, they enable the institution to document its performance. This can be extremely useful in documenting evidence for the NCA and the Baldrige criteria.

Quality award programs can also be coordinated with specialized accreditation reviews, as illustrated by the College of Business and Economics (CBE) at the University of Idaho (UI) (Lawrence and Dangerfield 2001). The university completed the accreditation process using a fresh approach that provided several important advantages compared to the more conventional accreditation review procedures. The CBE's method for the self-evaluation portion of the specialized reaffirmation of accreditation by the Association to Advance Collegiate Schools of Business (AACSB) was structured around the college's application for the Idaho Quality Award (IQA) in 1998. The IQA is a state award program founded on the lines of MBNQA, but is leveraged by the state of Idaho for encouraging improvements in their state organizations using total quality management techniques. The CBE believed that there is a large overlap in the procedures for applying for reaccreditation by the AACSB and applying for the state quality award. In addition, there are added benefits because applying for the state award adds value to the reaccreditation endeavor by providing the opportunity for establishing outside legitimacy with important college stakeholders and enabling the implementation of an established organizational learning structure to strengthen the reaffirmation of the specialized accreditation process.

However, it is important to identify the potential obstacles for colleges and universities that seek to use quality award programs in conjunction with accreditation or reaccreditation efforts. These obstacles include organizational identity, definition of quality, and unit of analysis (Seymour and Dhiman 1996). Organizational identity labels the often-common perspective of many individuals at colleges and universities who believe that their situation is too special for the usage of quality award processes that may have originated outside of higher education. The methods and language of management and business can present a barrier to nonbusiness faculty members and leaders who are not schooled in the area.

A second potential obstacle is defining quality. Higher education institutions often believe that they know quality because they see it, they are accredited, they have resources, and they are selective in admissions. These definitions may be internally acceptable quality definitions, but are obviously unacceptable in the context of a quality award system such as the MBNQA or a state award program such as the IQA.

The third obstacle is the unit of analysis. It is related to the highly decentralized nature of postsecondary education where many things happen at the departmental level. The departments in academe are loosely structured arrangements, whereas quality award systems assume or require that the organization consider itself as an integrated system. Despite these obstacles, and after reviewing the benefits and drawbacks of coordinating the application for a quality award with a reaccreditation process, an increasing number of institutions such as the UI decided that the prospective advantages outweighed the additional time and possible problems in implementing an innovative combined approach.

The preparation and crafting of the application documents for a quality improvement award are quite similar to an accreditation renewal in that the operation is complicated and involved. The strategy used by the CBE to address this effort was to appoint a faculty member, who was very familiar with the Baldrige criteria and the IQA, to assume the chief role in writing the application documents (Lawrence and Dangerfield 2001). The writing work was aided by the dean and two support staff members in the dean's office. Sections of the draft were distributed to all members of the college faculty for feedback and evaluation. Although the IQA has no special standards for educational institutions, the CBE applied the same criteria as businesses do within their state. However, as with the majority of business schools and many regular business organizations, the CBE has several interrelated yet separate product or service offerings, namely teaching, research, and service.

After reading the application, the IQA examiners identified several concerns in how the CBE was managing the research and service descriptions in the application. They also were required to clarify how the CBE was not a completely separate organization but a part of the larger university organization, akin to the subsidiaries and divisions of large corporations. Another concern that was identified by the quality award examiners was the issue of the relative independence of faculty and the concept of tenure. The examiners needed to be educated on the full definition of tenured faculty relationships with the college and correct misconceptions about tenure by individuals outside of higher education. The clarification included an explanation of the benefits of academic freedom in teaching and research, and also explained their post-tenure review every five years.

The participants in the quality award process at UI discovered that that there were significant benefits in the application process as well as in winning the Idaho Quality Award. It was reported that faculty members appeared more interested in the substance of the IQA application process than in the reaccreditation self-evaluation report for the specialized accreditation agency (Lawrence and Dangerfield 2001).

There was a good amount of learning by faculty members about the performance or nonperformance of other departments, as well as improvement information that the college gathered as a result of the IQA procedure for application. The feedback report from the quality award supplied the college with a helpful outside viewpoint that was special because of its thoroughness and wide-ranging organizational perspective. In addition, the college obtained additional valuable public recognition by receiving the award, presented by the governor at a formal ceremony conducted in collaboration with a statewide quality conference at the capital. For public institutions, these newspaper stories about success are very helpful as they seek resources from public coffers and market themselves to prospective students.

An alternative quality improvement strategy that is being tailored for accreditation programs by at least one regional accrediting association is the Academic Quality Improvement Project, built primarily on philosophy gained from the field of business management (AQIP 2002; Gose 2002). This improvement technique blends strategic planning tools with established accreditation approaches and facilitates increased coordination among postsecondary institutions and accrediting associations when compared to a more traditional comprehensive regional evaluation. There are a number of differences between the AQIP method and the Baldrige Award approach. There are nine criteria in the AQIP, that is two more than the seven measures for Baldrige. The results of each criterion are reported in the AQIP whereas Baldrige combines all the results. The AQIP procedure contains discrete criteria for various work practices such as teaching, partnering, administration, or research, while only the sixth Baldrige criterion addresses work methods. Additionally, the AQIP program is designed completely for postsecondary education as a section of an accreditation evaluation, typically a reaffirmation of accreditation. The US state-sponsored quality improvement programs and Baldrige, on which they were modeled, are award or recognition programs, not methods of accreditation approval. They therefore possess various objectives, and accordingly utilize approaches that may be seen by other stakeholders or those involved as not suitable for higher education accreditation.

The AQIP proposes that their strategy for improvement seeks to encourage ongoing institutional enhancement and verify the overall quality of higher education institutions to the public and external funding sources. The North Central Association accrediting agency has been using the AQIP approach for a number of years, and on the whole is relatively advanced in their approach to improve traditional accreditation strategies. The NCA–AQIP arrangement invites college and university representatives from participating institutions to be present at a three-day strategy forum to ascertain and cultivate three or four ambitious objectives that they hope to accomplish within the three-year time frame. The leaders at the institutions involved have reported that though AQIP actually entails a great deal more employee time, yet it is worth the extra time and expenditure. Toward the end of 2002, there were roughly 70 colleges and universities implementing an unconventional accreditation procedure offered by the NCA, and a good number of these are associate-degree granting community colleges.

The AQIP accreditation procedure involves four steps during which colleges and universities advance in a discovery process of interests and comprehensive assessment, that are both required initially. This is followed by a strategy forum to sharpen action projects and a systems portfolio, both fundamental to AQIP. These recur at three- to five-year intervals (AQIP 2002). The objective of the initial preliminary interest exploration phase is to ensure that the institution fully comprehends quality improvement strategies and the level of AQIP involvement, and is prepared to fully commit to or continue with a quality improvement strategy.

The second phase of the process is composed of a wide-ranging self-evaluation, where the institution conducts a quality-type formative analysis of its systems from an external viewpoint. Colleges and universities may select their method of self-evaluation. Institutions that are beginners to quality improvement strategies may choose an option that will enable them to conduct this self-evaluation in a timely manner and properly, with a comparatively reasonable budgetary investment of staff resources, employee time, and monetary expenditures. The prospective procedures include:

- A state quality award application and review (including either a feedback report or a site visit)
- A Malcolm Baldrige National Quality Award application and review (including a feedback report)
- An ISO 9000 registration application and audit
- A review process (using a quality system framework, such as the Baldrige Criteria, the state quality award criteria, the ISO 9000 standards, or the AQIP Quality Criteria) undertaken internally with the assistance of a consultant or other outsider familiar with quality principles and perspectives
- A Continuous Quality Improvement Network (CQIN) Trailblazer or Pacesetter Review (AQIP 2002, p. 668)

The strategy forum is a three-day affair, necessary for all new AQIP participants. It normally takes place near the NCA headquarters in Chicago (Gose 2002). The goal of the event is to enhance the action projects that have been created by the institution; and there is added value when representatives from several colleges and universities share their understanding and discuss their planned objectives. The representatives of the applying institutions are expected to attend a strategy forum every three to five years that typically corresponds to the time when postsecondary educational institutions are completing projects and determining the next set of objectives. The institutions participating in the AQIP are required to prepare a systems portfolio toward the end of the first three years. This document normally runs to 100 or less pages, and clarifies the major systems that the college or university use to achieve its mission. This portfolio shapes the structure of the reaffirmation of accreditation decisions, and is evaluated by quality-improvement authorities, involving at least one external expert.

The use of these quality improvement strategies for accreditation management can enable colleges and universities to institute a comprehensive accreditation strategy for overall organizational development, as well as enhancement of public viewpoints regarding the institution's performance and value. Nevertheless, these accreditation strategies should be thoroughly evaluated by postsecondary institutions prior to implementation and proper support should be garnered before it is begun (Bishop et al. 2001). According to the NCA–AQIP director Dr. Stephen Spangehl, the majority of institutions he has communicated with and have been performing organized quality improvement approaches, have notified him that it is a common problem of not putting enough institutional effort into the area of cultural shifting. That is, educating internal institutional participants to imagine in new ways, to contemplate terms of process, and to ponder working on inter- or intra-institutional committees so that a broad-based population leverages their different viewpoints to help solve common organizational problems that hamper accreditation efforts (Browne and Green 2001). These concerns and others are just one piece of the challenge that education leaders at higher education institutions need to think about, as requests and demands for greater answerability and higher value continue by the tuition-paying public, elective representatives, and other influential parties. In regard to the view held by many people that accreditation is the primary method by which colleges and universities can regulate themselves, and any alternative and perhaps superior methods are challenged by negligence, the quality improvement strategies examined in this chapter may provide some optimism for new, more significant, and perhaps enhanced self-evaluation methods for postsecondary institutions to scrutinize and advance their quality together with well-respected outside recognition.

It is wise for colleges and universities that consider adopting the innovative quality improvement strategies in coordination with accreditation management efforts to learn from the experience of others who have gone through this process. Some of the lessons learned include assigning someone at the institution who has experience with the quality award system to assume the chief role in preparing the quality award application (Lawrence and Dangerfield 2001). If none of the faculty or administrators have this experience, any member of the staff can be asked to volunteer as reviewers for national or state-sponsored award programs so that the institution can benefit from the training that is often provided. There are also a number of books and articles that provide explanations of how to understand and implement the quality award criteria (Blazey 1998; Blazey et al. 2003; Brown 1997; Caravatta 1997; Seymour 1996; Seymour and Dhiman 1996). It is also important for institutions to deliberate on the methods of handling the research component of institutional performance for the application, because some reviewers may expect a solid evaluation of the research customers, requirements and measures of satisfaction in relation to the mission under the stated division of the three primary issues of teaching, research, and service. In addition, institutions that have gone through these processes have reported that it is important to mention the critical role that support staff play in the application process, and proper acknowledgment should be documented (Lawrence and Dangerfield 2001). Finally, institutions must perform the work required to reformat the quality award application in the manner needed for the reaffirmation of accreditation self-evaluation report. This will assist those reviewers who may not be familiar with the quality award standards, and the proper wording in relation to the accreditation standards can go a long way in ensuring acceptance of the reaccreditation application. Accreditation reviews for initial and reaffirmation represent an important component of the overall quality assurance strategy for nearly all postsecondary institutions. By coordinating the application for a quality award into a reaccreditation process, the institution can add a beneficial feature to the process and supply the college or university with valuable advantages.

CHAPTER 19

New Electronic Tools to Support Accreditation Strategies

In addition to quality-improvement management strategies adapted from business, new software and Internet-based tools can be leveraged by colleges and universities in the accreditation process and beyond. Many institutions pursuing regional and specialized accreditation already use internal software for administration and teaching, and have now begun to use these tools for planning and implementing accreditation self-study efforts. For example, in the fall of 2002, the provost's office at Northern Illinois University (NIU) was confronted with a project to craft a self-study document for regional accreditation by the Higher Learning Commission of the North Central Association of Colleges and Schools (Askins 2003). The university sought to create a procedure that could effectively manage 12 committees of faculty members and administrators and a large amount of information. The institution knew that the committees were required to collect data, communicate information, coordinate the project workflow, and submit draft documents. The procedure created by the institution had specific requirements such as web-based information technology, secure access, and a dedicated subsystem for this purpose. In the light of these specifications, NIU decided to use their Blackboard electronic learning (e-learning) system, which is used for Internet-based course instruction. This system was already in use, and many faculty members and administrators already had good experience in using it and knew the system's abilities and weaknesses. Since Blackboard was ready to be used immediately, there was no need for customization or creation of special software; it could easily be modified by users for specific projects. Blackboard allows users a method to collect internal institutional information using special announcements and customized links to institutional data. It also has a tool to collect outside information using its external resources feature, various synchronous and asynchronous communication methods, easily customized communications to individuals and groups, and special subcommittee work areas on the website.

Since e-learning systems such as Blackboard were originally developed for online instruction, their educational features had to be adapted for use in the management of accreditation processes. At NIU, this involved "enrolling" several hundred faculty members and administrators in the new "course" that was the accreditation project (Askins 2003). Blackboard has several types of participation roles, including instructors of the online courses who oversee and manage web-based or web-enhanced courses as well as teaching assistants and students. For the accreditation process, NIU decided that high-level leaders such as the associate provost, the accreditation steering-committee chair, and the research associate who led the technical support would all be designated as instructors in Blackboard. The 12 important subcommittee chairs would all be teaching assistants, and a large number of subcommittee members would be enrolled as students. Other special individuals were also enrolled in the accreditation course according to the duties they performed. For example, the institution's president enrolled as a student so that he could visit the accreditation site at any time and remain updated on the progress of this critical project.

After enrolling the participants in the online e-learning system, the next step involved setting up a training schedule for the steering committee to learn the general features of Blackboard and specific details about the accreditation course structure. This was followed by virtual team meetings that were announced on the course calendar; they often used the virtual classroom for synchronous discussion or asynchronous discussion boards. Other team leaders preferred to communicate more simply using e-mail, which is also facilitated effectively by the Blackboard e-mail system. The discussion boards in Blackboard are somewhat similar to a listserv, with specified topic threads to be checked periodically by team members. Information management is greatly enhanced by the use of the Blackboard system. Subcommittee chairs were assisted in their need for vast quantities of university information in the crafting of their chapters of the self-study report. NIU leveraged two processes using Blackboard to assist information gathering. The first was the web address list area of Blackboard that was filled with web links to important university web pages useful to the accreditation project. This enabled the subcommittee chairs to efficiently locate important institutional information such as academic policies, procedures manuals, and other documents. The technology allowed quick adjustments and timely changes by the users, so that a useful web link could be shared with the accreditation team very quickly.

Aside from communication and managing data, another very useful feature of Blackboard is document management for the accreditation self-study. The Blackboard course-documents area was specifically reserved for unedited original data (including the preliminary drafts of the 12 self-study chapters for the accreditation report), and to function as an electronic version of the traditional resource room that many colleges and universities often set up for accreditation. The documents in the course-document area are the necessary documents that back up the findings presented in the self-evaluation report. The subcommittee chairs were directed on official listing procedures at the start of the endeavor. In the initial stages of document creation, the subcommittee chairs could use the Blackboard system for file exchange with their team members, the digital drop box, or the course assignments feature to assign specific duties to some team members or an entire group.

The group work area can be set up privately so that team members can be free to work securely, in confidence, and only open the work in progress to outside viewers when they so desire. Once the subcommittee teams have submitted their chapters, the steering committee chair can easily collate the various chapters into the final study using these tools. Therefore, the overall mission of using the Blackboard e-learning system was to store and work on the accreditation documents and to facilitate communication among members of the accreditation project. In addition, only peer-reviewers assigned by the regional accreditation agency can be given secure access to the Blackboard course website to see the raw, unedited information and other documents that are posted as evidence for the self-study. This provides a solid trail of the large amount of work and hours that are devoted to the accreditation review (Askins 2003).

Information technology is omnipresent in higher education today, and e-learning systems have become extensively implemented tools for distance education, corporate training, and enhancement of regular in-person or on-campus courses and programs. Growing commonalities among postsecondary educational institutions and business strategies for organizational development are widely recognized (Alstete 1996a), and consequently it is quite understandable that certain educational technology information systems can be used for goals other than those for which they were initially created, such as accreditation management and ongoing institutional improvement. The relatively new e-learning course information systems are an illustration of this development. They can be leveraged to facilitate individual employee and institutional learning (Alstete 2001). Learning by organizations (including colleges and universities) is more important than ever because of the very competitive postsecondary education market, and because managers must be able to discover opportunities, face problems, and pursue innovative ideas (Yeung et al. 1999), and subsequently convert those concepts into strategies that can used throughout an organization. The e-learning course systems are very suitable in this information era because administrators and faculty team members are or should be continually learning, particularly while completing accreditation self-studies. In addition, ongoing lifelong learning and using new technology should be encouraged, research abilities enhanced, and increased communication enabled for steering committees and subcommittees using any means available.

Accreditation committees can be more valuable than individual efforts, more proficient in writing output, possess a greater dedication to the duties because of team-peers, and finally produce a superior product (such as an accreditation report) (Natale et al. 1998). These enhancements are partially based on the improved results that committees and teams produce when they have an esprit de corps and affirmative synergy created by the harmonized energy of the accreditation team. It is well known that team efforts yield a higher level of execution than the total output of the individuals in an environment (Robbins 2000). E-learning systems can assist teams in achieving enhanced output for accreditation. Nevertheless, there are also potential difficulties in leveraging teams that accreditation management should be aware of. Management guru Tom Peters (1992) professes that although more subject knowledge is required and independence supported, there is also a requirement for accountability in the participation on project teams. Organizations such as colleges and universities should strategize on approaches that promote, recognize, and incentivize individuality and make possible effective team participation besides providing organization and answerability. Due to their academic foundations and related aspects, e-learning systems can provide accreditation committees with prospective answers to the team potential challenge posed by Peters.

Applying E-Learning Systems to Accreditation Management

The information technology tools that are available for use in today's e-learning systems are not truly new because they have been implemented occasionally in general team administration for at least 20 years. However, a benefit of these inclusive software packages is that organizations are not required to buy new costly hardware and software, as they may already be in use. Tools such as asynchronous discussion boards, e-mail, synchronous virtual chat, electronic document attachments, user digital drop boxes, external lists of Internet links, and other tools have been in use for many years at institutions of higher education. Accreditation committee members can conveniently use their home computer systems with Internet access to participate in the ongoing and time-consuming activities of an accreditation preparation. Committees of highly educated individuals at colleges and universities are not truly effective teams except when they are properly working together and mutually supportive (Robbins 2000), and features of the e-learning system technology such as those mentioned earlier are not a practical unified effort unless they are managed effectively. These kinds of information tools can be productive if their use is properly planned and organized to facilitate enhanced performance by their handiness and implementation.

There are, available today, several popular commercial e-learning systems, such as Blackboard, WebCT, FirstClass, IntraLearn, Prometheus, and College, as well as many internally developed systems at colleges and universities. The two most popular systems, Blackboard and WebCT, have recently announced a merger. This may lead to a greater similarity among systems to make them easier for users to learn. E-learning systems are becoming very common at postsecondary institutions and corporate training programs for enhancing traditional on-campus courses and for implementing complete virtual distance-learning programs and courses. These software programs are not very costly for entry-level models, are easy to learn, possess stable technology, can be upgraded easily, and are very useful for team members to use from their homes or offices. In the light of the NIU example (Alstete 2001), there are a number of important features of the e-learning systems that can be applied to enhance accreditation committee performance. These include:

- Committee Announcements. This is an excellent tool for general communication with accreditation committee members and others, as it allows them to post important topics, upcoming meeting schedules, relevant news, project assignment changes, forthcoming due dates, and other information. Steering committee and accreditation subcommittee chairs can use this conveniently in an asynchronous manner because announcement contents may posted at any time of day or night for others to then read at their convenience.

- Committee chair and committee member background information. The e-learning systems usually enable the posting of biographical information about registered participants, and this can include their position at the institution, previous experience with accreditation processes, educational background, special training, assigned committee duties, a photograph (which can be helpful at a large institution with many faculty and staff members), and other information that committee participants may find useful. This can be especially helpful for new college or university colleagues who may not yet be familiar with everyone at the institution and on the accreditation project. This feature is also advantageous to the accreditation chairs when they are delegating specific duties or creating subcommittees.
- "Course" (or in this case accreditation) Information. Even though the e-learning course website pages may use academic terms such as "course" or "instructors" in the title, the course information pages can be helpful for accreditation committees to list information about subcommittees such as the objectives for each, charges to the subcommittees, meeting times and locations, calendars of activities, and so on.
- "Course" (or in this case accreditation) Documents. The course documents featured on e-learning platforms are especially practical for posting critical documents, worksheet files, and other types of computer files relating to the ongoing duties of the committee. As mentioned in the example from a recently accredited institution, accreditation meeting agendas, meeting minutes, accreditation report drafts, spreadsheets, data files, electronic slide presentations, subcommittee or special research findings, and many other types of documents can be stored here for use by the committee participants.
- Assignments. This is clearly functional for accreditation committee leaders to assign projects, create subgroups, direct individuals with particular duties, as well as the general committee workflow communication.
- Communication Features. The e-learning software systems generally have a good range of communication choices for accreditation teams, such as e-mail, discussion boards, virtual "classrooms," member rosters, and subcommittee work areas.
- E-mail. The e-mail system in e-learning software normally enables individual selection of committee members or the opportunity to send a message to the entire team with one selection.
- Discussion Boards. The asynchronous discussion boards tools are probably the most helpful and frequently used option in e-learning systems for accreditation committee management. The discussion boards, which were originally created for course conversations among students and faculty, can be set up by the accreditation committee chairs and used by participants to write ongoing conversation threads anytime or anywhere there is Internet access. This allows accreditation teams to participate actively in discussions even if they do not attend a particular meeting on an important topic. Accreditation committee leaders can also establish general project-related discussion boards for broad committee member questions, as well as discussion boards that are specialized for the particular topic they are involved with or wish to communicate on.

- Virtual "Classroom." An online e-learning system virtual "classroom" is a dedicated live work area that allows (and requires) the accreditation committee members to be on their computers at the same time for a meeting so that they can participate in a real-time discussion. This synchronous feature is particularly useful when the committee needs to work quickly on accreditation decisions and project planning. In addition, many virtual classrooms in the e-learning systems mentioned here often have a "whiteboard" tool in their chat areas that enable the committee participants to collaborate visually in crafting figures, illustrations, sketches, and diagrams so that all members can see them and work together at the same time. The software also allows virtual slide presentations to be created and viewed. Another useful feature is that after a live classroom chat is completed, it has the option to be archived for future reference by other committee members who may not have been able to participate in the virtual meeting, by other steering committee chairs, or by the external peer-review team when they need to review the work of the accreditation steering committee.
- Roster. A seemingly simple yet very functional feature that is quite useful for all committee members is the roster, which enables easy and fast access to the names and e-mail addresses of other committee or subcommittee members.
- Group Area. Separate group work areas for subcommittees can be created by committee chairs for private discussion boards, separate e-mail lists, external links, and document digital drop boxes.
- External Link. The external Internet address links in both the subcommittee work area and the overall e-learning website viewable to all members can be established by the committee chairs and members to post accreditation links for internal and external web-based resources that are especially useful for the accreditation project. As Internet web address links are added, the committee chair can write a description about each site and explain why accreditation committee members should read it.
- Assessment. Because the e-learning systems were originally designed for course instruction, there actually are several tools that are available for committee chairs to use in assessing and managing individual and subcommittee activity. The leaders can track individual committee participation level and subcommittee performance using features such as the online gradebook, participant surveys, and quizzes to gather and report committee opinions. Other useful committee assessment statistics that report system usage by committee chairs include:

  - Total number of e-learning accesses in particular accreditation work site areas;
  - The number of individual accesses over time;
  - Participant visits to the website by hour of the day; and
  - Total accesses by individual committee members.

- Tasks (duties assigned by the accreditation committee chair), activity calendars, accreditation project schedules, and other management tools that assist the workflow and overall committee administration. (Alstete 2001)

All of these online e-learning software features can be combined to improve the performance of the accreditation committee by enhancing the ease of member communication and committee chair oversight. The issue that then arises relates to the types of accreditation committees that can be enhanced with this e-learning technology.

Leveraging Technology for Different Kinds of Accreditation Teams

Management research on team building provides information on various types of team (or committee in this case) formations and purposes. In regard to utilizing e-learning systems and virtual resource centers for accreditation management, the types of teams can initially be either entirely online (virtual) or technology enhanced. Virtual committee participants can work closely with one another on accreditation projects even though the faculty and staff members may have different work schedules, be separated by distance, or, in the case of large interstate or international institutions, be in completely different time zones in other parts of the country and world (Henry and Hartzler 1998). The e-learning systems illustrated here are particularly good for online accreditation committee work because the systems enable cross-functional committee participants to work on accreditation documents and research endeavors for a period of time (often directed by the committee chair) by using the variety of communication tools available in a rational and ordered manner. The relative ease of learning the systems, low initial cost (or no cost if the college or university already uses the e-learning system), and broad accessibility to the e-learning platforms from nearly anywhere make these new information systems very functional for accreditation management operations. The accreditation committee chairs and subcommittee chairs can plan the arrangement of these systems and closely supervise the individual and committee performance by utilizing the previously mentioned course assessment tools that are a part of most of these education-developed learning systems. Although virtual teams (and accreditation committees) usually have a finite duration (Bal and Gundy 1999), they could be permanent if the institution desires and if the committee continues to conduct important work for the college or university. The use of virtual committee technology by the accreditation committees will likely spur increased usage by other institution committees and generally increase the productivity and efficiency of many organizational operations and governance processes.

Aside from completely virtual online team usage, accreditation endeavors will most likely use e-learning or virtual resource centers to enhance regular activities. This normally occurs when an accreditation committee meets on-campus regularly, and then uses the e-learning system or other web tools as supplements for the basis of communication and ongoing operations. The features and advantages of the e-learning system that were explained earlier can actually become the foundation that individuals depend on highly for the accreditation committee work, inter- and intra-committee communication, and self-study report creation. Even though some people may see the e-learning platforms as an opportunity to do away with the requirement for on-campus accreditation meetings, there are still many individuals and leaders who believe that "true" committee meetings are only in-person and are essential for the delicate nuances, vigorous debate, and other such processes that can only occur in a traditional meeting. Improvement of the accreditation committee performance can be supported and managed by the committee and subcommittee chairs, and performed by the individual faculty, administration, and staff members only if they use and are fully supportive of the new information technology tools available in the e-learning systems. Therefore, it requires diplomacy, tact, and effective management techniques to obtain the buy-in and support of the highly educated, intelligent, yet sometimes obstinate characters encountered at colleges and universities today.

A potential challenge to enhancing committee performance in the e-learning system usage by established committees is encouraging and convincing the individuals to participate fully and actively in the usage of the accreditation online tools. The institution as a whole and the senior faculty leaders also face obstacles in convincing the subcommittee chairs to effectively use the e-learning systems and observe their use by their committee members. Senior administrators can support the usage and help overcome challenges by publicly admiring early adapters at the college or university who leverage e-learning systems for committee usage (because then those individuals who are not using it may be encouraged to catch up), by personally implementing the information technology for their own committees, and by exhibiting the potential enhancements of performance along with the ease of use.

Regardless of whether the accreditation committees are created just for the self-study portion of the accreditation endeavor (short-term) or are a part of ongoing long-term formations, and whether they employ completely online (virtual) e-learning systems or are just enhanced with such system usage, it should be understood that in general, accreditation committees and subcommittees that use this technology can be designated as problem-solving, self-managed, or cross-functional teams. In management studies, problem-solving teams or committees are formed by participants who share concepts or strategies to provide recommendations on how work procedures and activities can be improved in an organization (Robbins 2000). This is not unlike many accreditation-related committees and subcommittees that are set up for a self-study review. Additionally, regular project-focused teams often utilize various resources all through an organization to provide for a particular project (Yeung et al. 1999). For these project- or assignment-focused committees, the selection of e-learning software tools such as separate group discussion boards, document digital drop boxes, group or individual e-mail rosters, and special group external Internet link lists can allow them to remain attentive to the specific project and assist the chairs with supervision and follow-up documentation. The electronic digital structure of the online discussion board text messages and attachment features can also help participants and leaders to truly use the accreditation information that is provided and discussed. This makes the crafting of the self-study report drafts and final reports more efficient by including sections of or all the related documents as they may be needed.

Leaders can set up self-directed accreditation committees to use these e-learning systems and enable the benefits of using the technology to assist in the arrangement of committee duty schedules (by using the calendar tools), create specific performance objectives (by using the task assignment feature), interacting with other accreditation subcommittees under the institutional steering committee (by using e-mail and other communication features), choose individuals to work with, and evaluate committee performance. The independent characteristics of today's e-learning systems are very appropriate for the strategic objectives and operational processes of accreditation self-studies and maintenance of accreditation.

Accreditation cross-functional work committees are similar to management cross-functional teams that usually consist of employees who organize their labors to perform a task, are from roughly the same organizational position level, but are from different departmental work areas (Robbins 2000). These cross-functional committees at colleges and universities can also be composed of senior faculty and administrators from within a single school (within a larger university) who direct a number of departments. The cross-functional committees can even include members from other schools or colleges. In the corporate arena, the cross-functional team members can be from a particular company unit or other units, and may even contain suppliers and customers (Yeung et al. 1999). In higher education accreditation endeavors, this can involve using the information technology e-learning systems to include the external peer-review team members from other colleges and universities. A number of traditional organizational and geographical limitations that were conventional challenges for cross-functional teams can now be reduced or eliminated by properly managing the e-learning system technology tools that are available for accreditation management and general institutional ongoing usage.

Technology Enrichment and Improvement

These Internet e-learning systems can do more for higher education than merely improve individual team performance—information technology can also spur innovation. Through the application of information technology to restructure and reinvigorate team management, beyond just streamlining the functionality, new methods to support team efforts can be introduced (Stewart and Kleiner 1995). The important advancement of e-learning systems on committee or team general performance management focuses on the enhanced team spirit driven by the esprit de corps of the members, and reduces many of the particularly overbearing characters that may be present. Accreditation issues can be examined logically, communication with educational institutions is facilitated, and the large amount of in-depth data available on the Internet can be supplied to the team for informed decisions (see figure 8).

A top-down strategy for committee or team management is required if the approach is to become part of an institution's operation (Natale et al. 1998). This requires support from the senior administrators at institutions because without it, the usage of e-learning systems and new approaches using information technology will not work as well as it could. Information technology and the innovative use of it (such as alternative uses of e-learning systems for committee management) allow the senior administration to not merely support team efforts, but also to supervise and participate in these important activities.

One of the reasons that committee or team innovation using e-learning software systems is necessary today is that information technology and its use by many people have become a critical part of success in the information-rich modern era (Stewart and Kleiner 1995). The creation and ownership of knowledge and information are

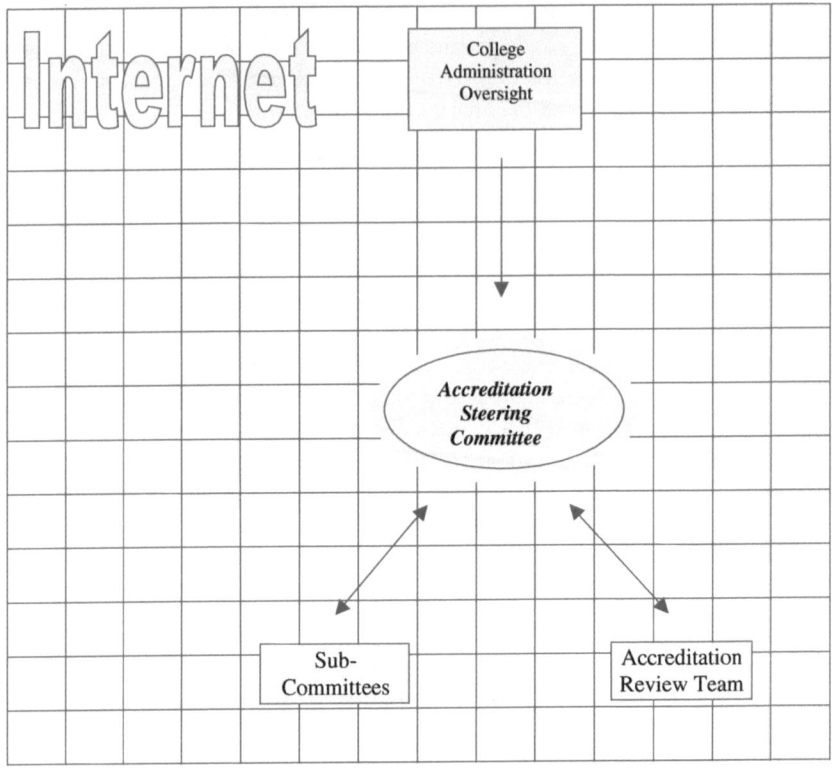

Figure 8 Internet Support of Accreditation Management

now the primary production or service forces that drive the economy, instead of traditional factors such as land, raw production material, or physical labor, as it was for many years (Bell 1973; Hoffman 1994). Creators and conveyer of knowledge, information, and training to support the economy and move knowledge forward are fundamentally important to higher education. The complicated and quite large information organizations in the corporate world as well as education today are especially appropriate to direct teams or committees to use e-learning systems for team enhancement because their duties often involve decoding rather complex organizational or topical problems, arriving at thoughtfully considered decisions, and employing imagination. Additionally, team participants in the information-rich modern era should possess a broad range of abilities and technology (Scott and Walker 1999). The e-learning software tools that are available today will probably evolve in coming years to include even better functionality and additional features. Accreditation committees at colleges and universities will come to depend even more on the information and communication abilities and the increasing rate of change of the information technology systems that are driving higher education institutions, enabling the committees to use the tools available to accomplish their assigned objectives.

However, putting team-based management practices into use is not enough assurance of impacting and improving organizational effectiveness (Elmuti 1997).

There is a need for developments in person–job associations as well as in the entire organizational structure. This requirement can be addressed to some extent with the utilization of e-learning systems by organizational teams. Even though the e-learning systems were not originally designed for noninstructional purposes, higher education institutions and their administration are effectively becoming learning organizations to flourish in the information age. Teamwork, information, and knowledge are significantly enhanced by the Internet's abilities and by related software systems that can organize and correspond powerfully and with ease. Individuals are the crucial elements to achievement in any institutional endeavor, and supplying individuals and teams with modern tools to succeed, together with backing and direction from senior administration, will help ensure greater success. Faculty leaders and senior administrators should set good examples by not only backing the use of e-learning systems for institutional committees, but also by using and participating in these activities. Frequently, this requires new participant training and ongoing education by administrators, faculty, and staff. In the end, many would find that it is a small cost to bear in return for the additional organizational enhancement, innovation, and achievement that can be obtained.

Other Approaches

Online e-learning course software is one method to leverage information technology in an accreditation self-study process, but other institutions have implemented a more comprehensive approach using a virtual resource center (Lindquist 2004, 2005; Masters 2004). Accreditation agencies usually require the self-study document and preparation for the peer-review team visit, but allow institutions to fashion their own self-study procedure and create their individual self-evaluation method. An essential element to plan for in nearly all self-study processes is how to supply the information to the working committees and subcommittees who will be analyzing the college or university with regard to the accreditation standards and crafting the various chapters and sections for the self-study report. Part of this procedure involves two methods to supply information that were examined earlier: having teams collect the information as and when required or designating specific personnel from an institutional research office to reply to the accreditation committee requests for data. However, these can be seen as reactive, instead of proactive, methods for information support in an accreditation self-evaluation that may involve many dozens or even hundreds of individuals in the preparation of the self-study report. Therefore, using information technology and a web-based virtual resource center (VRC) is a logical and efficient development by an institution to respond to these needs.

For example, this approach was used at Arizona State University (ASU) in the preparation for its comprehensive decennial review and on-campus site visit by an 11-person peer-review team in the spring of 2003 (Lindquist 2004). As mentioned earlier, many regional accreditation agencies, such as the Higher Learning Commission of the NCASC require the institution being reviewed to set up a resource room for the on-site evaluation team. The resource room must contain a particular inventory of materials such as institution reports that are referred to as the self-study document; minutes of the accreditation steering committee meetings; institution procedure and policy manuals; university catalogs; finance reports; master plans; official bylaws of faculty and staff associations; governance papers; and assorted institution handbooks. Institutions normally begin to assemble the information and documents that will be needed well before the visit by the evaluation team, and due to the proliferation and increased use of information technology, colleges and universities today often include online access to Internet-based resources. ASU had more than 200 individuals participating (similar to the previous illustration at NIU) in the crafting of the self-evaluation report. Its accreditation steering committee chose to establish an electronic data and communication center to assist the self-study procedure as part of its self-study plan.

The data and communication center is useful for both data (numerical information) and documents (primarily words) that need to be easily accessed by the accreditation committee members who are crafting the self-study and the external peer-review team that is visiting the institution. The accreditation steering committee at ASU formed a resource team at the institution and assigned it to prepare data and documents to assist the accreditation chapter teams, and data specialists from institutional departments such as student affairs, finance, and the library were chosen to be participants on this research time and supported the creation of the data and documents center (Lindquist 2005). The instructions given to the resource team included directions to provide legitimate, reliable, precise data and documents to the accreditation chapter teams at the beginning of the self-evaluation to aid the teams in concentrating on the analysis of information and not on collection. In addition, the working teams were not encouraged to create their own data and information strategies, as is so often the case in large multicommittee projects, which can cause difficulties later in the self-evaluation procedure. The data and documents center was established to be the central repository for the accreditation committees and visitation team.

ASU also created an in-house software designed by its information technology department to serve as an electronic bulletin board (similar to the discussion boards used in e-learning systems); this feature became the accreditation self-study online workroom. File folders were set up for broad access, and most of these folders were used to post and archive meeting minutes, collaborate on data and receive feedback, and most importantly to store the working drafts of the self-study chapters for the report. Thus, an electronic trail of work was left by the various teams and strongly supported the procedures that were outlined by the accreditation commission for the on-site visit. Other items that were stored in the online folders for the accreditation teams were the criteria for accreditation by the regional agency, self-study objectives and schedule of events, the prior ASU self-study report for the accreditation agency, Internet links to self-studies online at peer institutions, general team guidelines posted for instructions for data support, and instructions on writing the report chapters. An announcement board was set up to effectively manage the process. It helped keep the committee members updated on the self-study timeline benchmarks, forthcoming activities, and other events that were important (Lindquist 2005). These advantages are in line with management literature on using information technology to enhance team performance and should be strongly considered for implementation. It is important to remember that senior administration and faculty leader support and usage are necessary for success. There are examples of web-based systems being created for use by committees, but with little or no actual usage by members because there were no instructions for use, no expectations by leaders in this regard, nor required usage.

At ASU, the institution implemented the system in the early stages of the accreditation process, and the subcommittee members were given data with easy and immediate access to help initiate their research tasks quickly without lengthy research that can often delay the start of a large writing project. The team members found that the advantage of using the virtual resource center was that they had convenient access to the criteria for accreditation by the association. In crafting the report, the teams were looking at the same data that helped ensure the use of approved institutional information for consistency throughout the very comprehensive and extensive chapters contained in a regional accreditation report. In addition, when the external peer-review team was preparing to visit the institution, they were able to visit the online information center and ponder on areas of deeper examination. It requested that specific materials be printed on paper to be filed in the traditional accreditation resource room. This saved the institution considerable time and space needed for their on-campus regularly planned resource room for accreditation. Visiting teams and institutional community members can continue using the information stored on the system after the visit for the necessary follow-up and response work that must be performed by both the institution being reviewed and the accreditation agency.

CHAPTER 20

Accreditation of Distance Education

In recent years there has been a large increase in the number of distance education institutions, programs and courses in higher education via distance learning by traditional colleges and universities, and overall student enrollment in distance education (Bower and Hardy 2005; Carneval 2004; Watts 2003). As with any large-scale change and especially with technology, there are challenges and concerns here as well. There is currently a level of divisiveness among many faculty members who feel pressured to produce distance education courses and those who believe that there is no excuse to not use the new electronic learning tools that are now available (Watts 2003). Those faculty members who are frustrated about not being able to fully use distance education might agree that while the reach to students is extended, the certainty of results is not assured. This uncertainty extends to many administrators, legislators, parents, students, and other stakeholders of higher education. Nevertheless, the recent explosion in distance education enrollment is projected to increase in the coming years, and this creates a demand for many institutions of higher education to seek assistance and guidance in managing the rising student populations for online courses (Carneval 2004). Fortunately, many regional, specialized, and national accreditation associations in the United States are addressing the challenges presented by the new era of distance education. The accreditation agencies to be examined here are providing excellent guidance on how to ensure that quality education courses and programs are being delivered.

The president of the Council for Higher Education Accreditation (CHEA) has written about the impact of distance learning and the appearance of new types of providers of higher education (Eaton 2002). The typology includes the following classifications:

- New freestanding, degree-granting online institutions: A small number of high-profile new providers of distance learning—sometimes called "virtual universities." These include degree-granting, nonprofit institutions and degree-granting, for-profit distance-learning providers.

- Degree-granting consortia: A network of institutions from which students may select a range of online courses and programs and earn a degree granted by the consortia.
- Nondegree-granting consortia: A network of degree-granting institutions from which students may select a range of online courses and programs but which require that students earn a degree from a member institution.
- Corporate universities: Corporations that maintain private teaching and training enterprises initially enrolling employees and, increasingly, enrolling outside customers as well. Many of these corporate universities are still site-based, but they are moving quickly to online modes of operation.
- Unaffiliated providers of online programs and courses: Online courses and programs that are not affiliated with any institution. These range from credit-bearing educational activities to single-instance noncredit offerings (for example, a four-hour online seminar). (Eaton 2002, p. 3)

However, the report adds that a hybrid model of distance learning combined with campus-based traditional teaching will become increasingly common in the near future. In this hybrid teaching strategy, conventional site-based and distance learning together with regular student support services will be offered in conjunction, whether from traditional institutions or the new provider-type of institutions.

Other recent reports have stated that between 2.3 and 5 million people completed some type of online course in 2004, and at least two-thirds of colleges today that offer traditional on-campus courses also offer online distance-learning courses (Allen and Seaman 2005; DETC 2004). The course and program offerings in online education have indeed entered the mainstream of higher education, and online higher education in both size and breadth of courses and programs is penetrating traditional institutions. The large majority of higher education institutions report that they are using core faculty to teach online courses, and the overall percentage of colleges and university that state online higher education is critical for long-term strategy is also in the majority. The largest increases are seen at two-year institutions, where 72 percent now agree that online education is critical for their continued success (Allen and Seaman 2005; Bower and Hardy 2005). Even though smaller schools, private nonprofit institutions, and baccalaureate colleges remain somewhat less likely to believe that distance learning is part of their long-term strategy, the sheer number of online courses and programs that these institutions are offering now makes this topic very important to consider when planning a management strategy for the achievement and maintenance of regional and specialized accreditation. The eight regional associations, the many specialized associations, and even a CHEA-approved national association dedicated to distance learning have specific guidelines that institutions should consider. Since this book is primarily aimed at the majority of institutions pursuing regional and specialized accreditations, these association guidelines and best practices will be examined further.

Regional accreditation agencies have found it necessary to stay current on developments in distance learning, in particular the monitoring of academic quality (El-Khawas 2001). There were, and still are, some lingering questions such as whether new accreditation standards were/are needed, how can programs that are delivered via distance learning be evaluated properly, how can the monitoring be integrated into regular quality assurance processes, and which of the many agencies should oversee the monitoring. The latest developments in the field of accreditation policy seem to have answered some of these questions, in particular the integration of distance-learning standards into most, if not all, accreditation standards by the regional and specialized agencies.

The recent challenges presented to higher education by the reauthorization of the higher education act by the United States legislators also involved some concerns about quality assurance, particularly with regard to the for-profit postsecondary institutions that are seeking to obtain federal student loan/grant funds based on their distance-learning programs. Media attention was drawn to new institutions that were specifically formed to leverage the new distance-learning technology, and there were critics of these profit-making entities who embrace the new online learning on a large scale. In particular, the University of Phoenix is a large online university with a distinctive academic approach, although a large portion of its current programs is also offered on campus-based locations. This institution is accredited by a regional association, and has been seeking additional prestigious specialized accreditation for one or more of its programs. The specialized accreditation agencies have been somewhat reluctant to adapt their traditional standards for such a radical departure from various input-driven measures such as full-time versus part-time faculty expectations as well as academic qualifications of the faculty. However, the eight regional accreditation associations have accredited a number of full distance-learning postsecondary institutions, as well as a plethora of traditional colleges and universities that now offer fully online degrees, programs, and courses.

The eight regional accrediting commissions prepared a "Statement of Commitment for the Evaluation of Electronically Offered Degree and Certificate Programs" (NCA 2001). The agencies realized that technologically communicated instructions offered to students at a distance have quickly become an influential element of higher education. As mentioned earlier, increasing numbers of postsecondary institutions are now offering online courses, programs, and entire degrees ranging from baccalaureate through doctoral programs, while the number of offerings is continuing to increase each year. To complicate matters, there are also new online educational providers that often do not possess the characteristics of an established college or university. While this trend is building opportunities to serve new student markets and (perhaps) improve service to current student groups as well as forcibly motivating innovation in traditional institutions, these new educational instructional methods challenge conventional quality premises, raising new questions with regard to the true nature and substance of a postsecondary educational experience for students and the resources that are needed (NCA 2001). In this way, the new online competitors and existing providers are presenting unexpected and noticeable trials for the eight regional accrediting commissions, as well as other accrediting bodies that are supposed to guarantee the educational quality of the large majority of degree-granting higher education institutions in the United States and other nations.

The regional accreditation commissions have created a strategy to address these developing electronic approaches to learning. It is articulated in a set of "commitments" that are designed to help guarantee the quality of distance learning that the public and others expect these agencies to perform (NCA 2001). The commitments include support for educational traditions, values, and standards that have directed the regional agencies' strategy to improvement of education; interagency cooperation that is designed toward a reliable strategy in the assessment and quality assurance guidelines of distance education through knowledgeable cooperation; and generally sustaining good performance among the institutions that offer distance learning.

The first commitment by the accrediting agencies to traditions, values, and principles is based, at least partially, on the more than 100-year-old history of US regional accreditation that has often included adaptation to a changing educational environment (as shown in the periods of accreditation history in the first part of this book) and on sustaining high-quality educational standards while also acknowledging diversity in delivery methods. Accountable developments have been supported within an accreditation system of responsibility rooted in long-standing ideals and standards through which educational quality has been characterized. The product of these efforts over the years has been an increasingly growth-oriented model of postsecondary educational offerings for the public that are characterized by an excellent range of offerings and high quality that continue to address the changing requirements of the populace. The statement by the eight regional associations—an impressive document that espouses forthright beliefs in quality assurance—shows that the agencies are truly seeking to respond properly to the new forms of distance education that are likely to have an ongoing and institution- and industry-changing effect on academe. The agencies acknowledge that this will be an ongoing effort, because educational change continues together with information technology developments. This makes any strategy to establish permanent definitions of the primary foundations and conditions in distance education, as well as the standards to apply, neither achievable nor even required (NCA 2001). Instead, the regional agencies' response to distance learning will change incrementally over time. However their success in distance-learning accreditation so far shows an excellent ability in quality assurance by the individual regional accreditation associations that have quickly adapted to the strong inventiveness of higher education institutions in greatly increasing the use of information technology to improve and expand their educational offerings.

As the accreditation agencies continue their evolution of the quality assurance and improvement strategies for higher education programs that are delivered via distance learning, the regional agencies will carry on with their goals to obtain a dual commitment to accountability and innovation. Accountability in the sense of external public recognition and assurance of quality, and innovation with regard to internal revitalization are concepts in line with the established goals of accreditation and the institutions that are dependent on the accreditation system. Colleges and universities, as well as accreditation agencies are seeking to maintain a balance between satisfying the requirements that regional agencies have as a trustworthy measure of institutional quality and supporting insightful and creative postsecondary developments. Educational leaders, faculty, and the public should be made aware that sensible divergence from conventional educational processes will be authorized, and those that are not of good quality will not be endorsed (NCA 2001). Recent data from the National Center for Education Statistics (NCES) reveal that public two-year colleges account for nearly one-half of all institutions in the United States that offer distance education (Bower and Hardy 2005). These institutions are quickly responding to the mounting needs of non-traditional learners (individuals outside the conventional 18- to 23-year-old student characteristics), including professionals employed full time and lifelong learners. Most, if not all, postsecondary institutions must learn now to compete effectively with many online educational providers and simultaneously address the demands by the public and legislators to rein in cost and continue with quality improvement.

As explained in the various accreditation standards, the regional agencies define institutional quality based on the mission of the colleges and universities being reviewed. The applicant institution will be approved for the important regional designation if its intentions are suitable for postsecondary education, if it has the resources required to achieve those intentions or goals, if it can validate that it is accomplishing them, and if it has the wherewithal to carry on successfully. This unreservedly adaptable standard is especially suitable for the evaluations of new methods for delivery of distance learning as well as electronic-learning enhancement of regular on-campus educational operations.

Therefore, the regional accreditation associations appear committed to maintaining specific values while also seeking to continue their dual approach of steadiness with elasticity in the assessment of distance-learning activities (NCA 2001). These commitments include the belief that online postsecondary education is properly performed within a community of learning where knowledgeable education professionals are dynamically and willingly occupied with developing, establishing, and improving the distance-learning educational programs. Higher education practitioners have been discovering that collaborating online provides useful direction for faculty members who are seeking to guide their students to work together in innovative assignments, and develop beyond the use of traditional writing assignments and lectures, while deepening the postsecondary learning experience for all through their facilitative online strategies within a collective endeavor (Bender 2003; Palloff and Pratt 2004). Community and online collaboration brings students and faculty closer in support of the education of individual members of particular groups, while also encouraging creativity and critical thinking. The accreditation agencies are also committed to promoting the philosophy of higher education in general that believes effective learning is vibrant and interactive, regardless of the setting in which it occurs.

The first commitment of the regional agencies to traditions, values, and principles also includes the intention to ensure that higher education programs leading to collegiate-level degrees possess integrity and are prepared on the basis of their substantive and rational curricula that delineate specific student-learning outcomes (NCA 2001). Distance-learning programs should be set up with well-planned and multilevel assessment strategies, based on the mission of the institution and program, and can be well documented for accreditation review. Colleges and universities should also recognize the requirement to deliver services regarding online student needs that are associated with the resources needed for their successful achievement in academic areas. The accreditation associations appear to be leaving significant leeway to institutions in this approach, and accreditation managers should use their creativity, judgment, and previous experience in addressing these expectations. Institutions pursing accreditation are responsible for the distance-learning education provided in their name, and if any portion or all of the educational service is delivered by third-party providers, the institutions must ensure that accreditation expectations and standards are being met. Student learning is the particular emphasis of the regional agencies in recent years, and institutions that undertake the assessment and improvement of student-learning quality must pay extra attention to this expectation.

It is well known in academe that there are obstacles to overcome in supporting these critical beliefs in educational philosophy through distance education. The regional accreditation associations appear to understand this situation, and have identified specific values that may be articulated in authoritative new approaches as innovative colleges and universities are leveraging the new technology to stay current and accomplish their missions. The regional agencies will probably continue to restrict their range of authority to higher education institutions that grant degrees. Nevertheless, the associations are aware that a large number of the educational services that are offered using distance learning do not currently lead to college degrees, and instead abbreviate the length of the course and offer very specialized programs that supply specific training courses that result in various kinds of certificates for students who complete them. These increasingly common offerings by regionally accredited colleges and universities, or institutions that are pursuing regional accreditation, or even offered using the institution's name by a third party, are officially deemed as inclusive within the domain of the accredited institution and therefore subject to agency evaluation (NCA 2001). Program developers and accreditation administrators at colleges and universities should be careful when considering additional non-credit certificate programs via distance learning. This is stated not to restrict the current and future offerings, merely to provide high-quality offerings containing measurable student-learning outcomes that can be documented for the regional accreditation associations.

The regional accreditation agencies are aware that the educational concepts and practices that are progressively more associated with unconventional organizations and structures challenge traditional ideas of higher education with regard to what represents postsecondary institutions. Creating an environment for colleges and universities to use innovative cooperation and the implementation of distance learning has created extraordinary inter-institutional arrangements and relationship among regionally accredited higher education institutions, as well as with organizations outside academe. Some examples of this include the Western Cooperative for Educational Telecommunications (WCET) at the Western Interstate Commission for Higher Education (WICHE), The Learning House in Kentucky (which develops online courses and programs for colleges and universities), the American Distance Education Consortium, and other cooperative arrangements among institutions (Carneval 2004; El-Khawas 2001). This approach can, in some ways, help facilitate the meeting of regional accreditation standards and implement overall quality assurance, especially for small institutions that may be new to distance education. While commonly producing a favorable growth of postsecondary educational opportunities, an improved management of institutional resources, and inter-reliance on accreditation approval, these interorganizational cooperatives can also frequently result in a dispersion of accountability for the programmatic or degree quality in regard to student learning and academic achievement. Additionally, the quality oversight in these cooperative arrangements for distance education frequently relies on the ongoing availability of various institutional resources that are insecurely tied together (NCA 2001). The eight regional accreditation agencies, as they evaluate these cooperative partnerships, have stated that they will deem it necessary to ensure that accountability be explicitly determined and articulately detailed within the institution(s) that are regionally accredited, and that sensible assurances are supplied to help guarantee the ongoing accessibility and usage of the required distance-learning tools that are external to the particular college or university's jurisdiction.

The second commitment that the regional accrediting agencies have made with regard to distance education is on cooperation, consistency, and collaboration among the associations. The regional strategy to academic quality assurance has functioned very well, despite the critics, and despite their sometimes-wide differences in the markets they serve, the regional agencies are fundamentally similar. The eight agencies have leveraged an ability to reflect the large cultural diversity in the United States and this is expressed in their accreditation standards as well as their operations. The associations have implemented several examples of local (regional) postsecondary programmatic trials that have resulted in benefits for higher education in general. Additionally, the somewhat enigmatic and decentralized approach to regional accreditation in the United States has significantly encouraged the continuation of self-governance by maintaining a strong closeness to the members by the associations (NCA 2001). A large centralized postsecondary educational bureaucracy would probably not be as successful, innovative, or encouraging of new developments such as is currently being experienced with distance education.

Technologically or electronically delivered learning programs, increasingly asynchronous (not live or real-time) and based on the Internet, and therefore not tied closely to a specific region or location, does create some interesting dilemmas regarding the appropriateness of the decentralized regional accreditation strategy to quality assurance (NCA 2001). Even though regional agencies acknowledge this paradox, they also understand that the large majority of higher education programs delivered in the United States currently is provided on traditional ground-based campuses, and that practically all distance-learning education programs resulting in college degrees are being offered by conventional colleges and universities that have significant academic structures within a single region that also needs to be evaluated by local peers. It is understandable that this situation may evolve in the coming years, but at present the regional arrangement for oversight of collegiate programs seems to be fittingly responsive to the existing realities of higher education in the United States and is actually quite successful in satisfying the requirements for postsecondary educational quality assurance demands by society.

Nevertheless, since the distance education methodology is turning into a very important factor in higher education, with many colleges and universities creating countrywide and international student enrollment markets to take advantage of online registration increases, the regional accreditation agencies in the United States have committed themselves to a large effort of programmatic degree consistency across the regions of the country, all the while being very congruent with their long-standing association sovereignty and self-governance in measuring educational endeavors. Furthermore, the agencies are striving to help guarantee that distance-learning programs and degrees offered by numerous institutions in various regions adhere to similar high-quality assurance accreditation standards by implementing a comprehensive appraisal structure that uses peer-evaluations that are similar in all the regions of the country. These standards include the concept that institutions involved in the initial development of distance education courses, programs and/or degrees that are created or available to students in noninstitutional and not nearby locations will be created with thoughtful and deliberate evaluation. As mentioned earlier, some established institutions in the United States such as the University College at the University of Maryland have moved to offer programs internationally (El-Khawas 2001). The university started with the advantage of having a large amount of experience providing collegiate instruction globally because of a contract with the US military, and is now enrolling a large number of students using distance-learning technology. Other accredited, long-established, and respected universities in the United States have partnered with commercial firms to offer distance-learning courses and programs, and even enacted organizational changes to meet the expected demand for distance education. Institutions such as Columbia University, Cornell University, and New York University have created special profit-making divisions to provide online distance learning. Some of these university divisions have not been completely successful, and a recent report states that the boom in distance education has not (yet) lived up to its pledge of fundamentally changing the traditional instructional environment and generating much additional revenue for higher education (Massy and Zemsky 2004). Although there has been some debate about that issue (Massy and Zemsky 2004; Simonson 2004), the number of enrollments and institutions providing distance education is certainly formidable.

Other commonsense accreditation issues on this matter include the strategy that postsecondary institutional efficiency in offering distance-learning programs will be clearly and scrupulously evaluated in conjunction with the regular periodic reviews of higher education institutions, as in comprehensive visits and required interim reports expected by the agencies (NCA 2001). The accreditation associations also expect that institutions will include self-evaluation as a required component in all accreditation review procedures to help ensure quality improvement, and in situations where insufficiencies are recognized in this area, or where there are problems regarding reliability, proper solutions will be planned and meticulously supervised. In cases where a college or university is found to be evidently unable to properly provide quality distance education offerings, suitable action by the agency that adheres to the specific regional association policies will be implemented. The accreditation agencies are implementing overall quality oversight of distance learning with deference to the regional criteria and local customs therein. College and university leaders should, therefore, be sure to have a thorough understanding of their regional accreditation expectations in general, as well as be up-to-date on distance-learning quality assurance when an institution plans to implement such a program or degree.

As each of the eight regional accreditation associations carry on with their development of monitoring the quality of distance-education programs, they appear committed to learning from the broad base of knowledge that they have accumulated together and are working together to address the challenges of new information technology as creative teaching strategies are implemented by colleges and universities.

Specialized agencies such as the AACSB-International have also developed statements identifying key quality issues in distance education, as have other specialized and national accreditation associations. While most higher education institutions offering distance-learning programs are primarily located in one of the six regions of the United States (as noted in Part One of this book, there are eight agencies because two agencies have subdivisions), thereby gathering them under the authority of one regional agency, the distance-learning technology is proving that aspects of completely virtual higher institutions are not confined to any one of the regions. In those instances, it is not clear to the institutions or the public at large which official U. S. regional accreditation agency should have the authority and accountability for educational quality assurance and oversight. Since there are still relatively few of these institutions, compared to the number of traditional land-based colleges and universities offering distance-learning programs or courses, without clear regional location this situation is just beginning to present a problem for the regional accreditations and is to be addressed by the Council of Regional Accrediting Commissions (C-RAC) through its mechanism for interregional protocols (CRAC 1999; NCA 2001).

It is clear that the regional accrediting commissions are working on their abilities to cooperate when appropriate such as in the case of distance learning, because the needs that arise for this reach beyond individual regions of the country. Also other authorities, such as the state education departments and the federal government, have a large stake in assuring the public that there is proper educational quality assurance overseeing distance-education programming. Based on the long experience of collaboration with individual state-based higher education offices and the United States Department of Education (USDE), the eight regional commissions have stated that they are undertaking a firm promise to perform their duties solely and collaboratively with the appropriate agencies because they have common objectives in assuring the quality of distance-learning education programs. The regional agencies will also seek the ongoing guidance of the Council on Higher Education Accreditation as an organizer and manager.

As part of their commitment to oversee distance learning, the regional agencies have also elaborated on their proper support of good practice in this area. As the field of higher education offerings continues to grow with educational opportunities through distance learning, the regional accreditation associations are committed to supporting good practice in distance education among associated colleges and universities. The agencies believe that this approach will support individual institutional improvement, and is in line with their mission to facilitate educational excellence. With this in mind, several years ago the regional agencies created and enacted a joint statement on the "Principles of Good Practice in Electronically Offered Academic Degree and Certificate Programs," developed by the Western Cooperative for Educational Telecommunications. This helped to strongly encourage the cooperative strategy to distance education among the agencies. The WCET was founded by the Western Interstate Commission for Higher Education in 1989, and is a membership-supported organization designed for providers and users of educational telecommunications (WCET 2006). Members represent the higher education community, nonprofit organizations, schools, and corporations. The WCET's membership is international; however the majority is in the United States. The agency claims to represent the very creative thinkers in the use of educational technologies for educational institutions, either on- or off-campus. In a follow-up to this development and in seeking to complement these distance-learning strategies, the groups of regional accrediting commissions through their national organization (C-RAC), arranged with the WCET to establish a more detailed explanation of those attributes that illustrate quality in distance education. Institutions that offer distance-learning programs and pursue accreditation should seek the expertise base of organizations such as the WCET and the significant experience of the regional commissions in assessing distance education. Their statement on "Best Practices for Electronically Offered Degree and Certificate Programs," offers colleges and universities a thorough and challenging look at what is considered current best practice in distance education today. It is currently being used to some extent by each of the regional commissions, and is harmonious with their regularly established educational policies and procedures to encourage and monitor continued good practice in distance education among associated institutions of higher education.

With regard to the specific guidelines for distance education that institutions should follow, in the example here, the regional accrediting associations have agreed upon the following definition and guidelines, based on an extension of the principles developed by the Western Interstate Commission on Higher Education (NCA 2006). Distance education is defined, for the purposes of accreditation review, as a formal educational process in which the majority of the instruction occurs when student and instructor are not in the same place. Instruction may be synchronous (live or real-time) or asynchronous (such as using e-mail or on discussion boards). Distance education may employ correspondence study (via the mail as in home study courses), or audio, video (video-teleconferencing is in use at many institutions), or computer technologies (the current form that is very popular). Any institution offering distance education is expected to meet the requirements of its own regional accrediting body, and be guided by the Western Interstate Commission for Higher Education Principles. In addition, an institution is expected to address, in its accreditation self-studies and/or proposals for institutional change, a series of expectations (to be elaborated further) that it can anticipate will be reviewed by its regional accrediting commission (NCA 2006).

In the important area of curriculum and instruction (a common concern among those new to distance learning), programs should provide for timely and appropriate interaction between students and faculty, and among students in the courses and programs. The principles state that an institution's faculty members assume responsibility for and exercise oversight over distance education, ensuring both the rigor of programs and the quality of instruction. This may be unclear about proper adherence by those institutions that outsource the service through cooperative arrangements, commercial firms, or have in-house departments to oversee the detailed content of online courses and programs. However, issues regarding oversight can be addressed if the faculty does indeed have overall responsibility. The institution should also ensure that the technology used is appropriate to the nature and objectives of the programs, and the currency of materials, programs, and courses. Course materials can easily be updated when using the online e-learning systems. This can be done by faculty members directly or through educational content from third-party providers. The key is to ensure that the faculty members have decision-making rights on the inclusion and exclusion of educational material.

Related to this area is the issue of the institution's distance education policies concerning ownership of materials, faculty compensation, copyright issues, and the utilization of revenue derived from the creation and production of software, telecourses, or other media products. Regional and specialized accreditation policies are becoming very specific with regard to this topic of intellectual property, and there has been much debate about intellectual property in recent years (CHEA 2000, p. 891; Claerhout 2000, p. 893; Sanders 2002, p. 892). In addition, colleges and universities should provide appropriate faculty support services specifically related to distance education, and include intellectual property policies as part of the institutional support. The institution should provide appropriate training for faculty who teach in distance education programs, and accreditation agencies will likely wish to see a schedule of training sessions that were offered.

Evaluation and assessment are a primary component of accreditation reviews. Institutions should assess student capability to succeed in distance-education programs and apply this information to the institutional admission and recruitment policies and decisions. The evaluation of admission processes can be screened in a variety of ways, such as technology-specific pre-entry requirements, stated policies, technology-mediated application processes, and standardized testing, along with other methods suggested by recently accredited institutions with distance learning or from experienced accreditation consultant evaluators. Benchmarking successful accreditation reviews from other recently accredited institutions can be done quite simply through Internet research, requests to colleagues at other institutions, and attendance at academic conferences or accreditation association meetings. Institutions should also evaluate the overall educational effectiveness of its distance education programs (including assessments of student-learning outcomes, student enrollment retention, and perceived student satisfaction) to ensure comparability to traditional campus-based programs at the college or university seeking accreditation approval. In addition, colleges should ensure the integrity of student work via distance learning and the credibility of the degrees and credits it awards. This can be done through a variety of methods, such as using multiple measures of learning in online courses with various tools including testing, writing assignments, discussion board evaluations, random cross-checking of a student's prior performance with online achievement, and correlation of individual student software usage statistics. One method that is gaining increasing popularity is the supplementary use of plagiarism detection software (Atkins and Nelson 2001; Baron and Crooks 2005; Foster 2002).

A common challenge for full distance-learning virtual institutions for both state-level and accreditation-level policy adherence, are the library and learning resources (NCA 2006). Traditional and established institutions will normally find it more easy to ensure that students have access to and proper use of appropriate library resources than virtual colleges. However, as the availability of full text online documents become common, the library requirement may evolve to allow for more modern interpretations of library expectations. In addition, institutions are expected to provide collegiate-level laboratories, facilities, and equipment appropriate to the courses or programs and monitor whether students make appropriate use of learning resources. These expectations are all closely related to student service standards of the accreditation agencies. Colleges that provide distance learning should also facilitate adequate access to the range of more or less traditional student services that suitably support the distance education programs, including student admissions, financial aid support, academic advisement, delivery of course materials (as defined by the type of electronic instruction system), and college graduate employment counseling and placement. The distance-learning program technology can assist the provision of this wide range of services, and institutions seeking accreditation should view the e-learning technology as a tool to leverage and not as a hindrance.

Colleges offering online programs should also be able to show that they provide sufficient protocols for resolving student complaints (NCA 2006). Institutions must be able to provide students with appropriate advertising, recruiting, and admission information that satisfactorily and correctly represent the degree programs, requirements, and student services that are available from the institution for distance learning. Additionally, colleges and universities should ensure that students who are admitted possess the knowledge and equipment necessary to use the technology employed in the program, and provide aid to students who find it difficult to use the required technology. This may be addressed by various grants, loans, and special need-based initiatives from internal and external sources. In the area of collegiate facilities and finances, the institutions seeking accreditation should possess the appropriate equipment and technical expertise required for distance education. Often, accreditation agencies will allow applicant schools to have a plan in place for certain accreditation standards, but for a requirement such as this for a currently active distance-learning program, it would be best to show existing expertise in use that is monitored carefully with assessment strategies. Finally, an institution's long-range strategic planning, budgeting, and policy development processes should reflect the facilities, staffing, equipment, and other resources necessary for the success of the distance-education program of the college or university seeking accreditation.

CHAPTER 21

Conclusion

As illustrated during this inspection of accreditation and its management, the relatively new business management quality strategies and coordinated accreditation approaches all necessitate robust leadership from senior administrators and high-ranking faculty leaders in colleges and universities. Successful strategic-level management of accreditation efforts are required for guiding postsecondary institutions to participate properly in the procedure of creating innovative accreditation policies, turning these policies into actuality in the future, and providing an ongoing awareness to the internal and external stakeholders about the achievements and possibilities for the future. The character and personality of the academic leaders are very important for effectively recruiting and designating the internal and external accreditation committee participants, and leaders, as well as establishing a large base of trust required for the guidance to be directed over the lengthy accreditation time period. The controllers of the accreditation self-study committees should be persons with good foresight of where the colleges or university should be heading, in addition to having an excellent understanding of the institutional mission and history. A report on leadership in higher education by Moore and Diamond (2000) proposes that the fundamental work of leadership involves the ability to appoint human capability in the quest of common cause. In the area of accreditation strategy, this common cause can include a development of the belief by the institution's administration, faculty members, staff, students, and board members that an innovative accreditation endeavor management strategy can achieve more than a mere official recognition from an agency. The strategy can also be a procedure that revitalizes the institution, establishes new objectives, and alters the viewpoints of key people. College and university leaders can positively involve faculty members and staff personnel with outside groups in producing and renovating an enhanced vision for the institution through commitment, connecting with required partners, and compelling management strategies that transform the vision into reality.

There are many obstacles that leaders of higher education face in the process of achieving their accreditation needs and goals. One considerable challenge is planning, taking into account the various and frequently conflicting cultures within the academy (Newton 1992). These more influential groups include the corporate community and the community of scholars. These quite different cultures possess ostensibly contradictory goals, yet they also have a mutually dependent relationship because they are equally necessary for the institution and their individual group to achieve success. Normally, the provost/chief academic officer of the college or university should guide the accreditation effort by arranging backing from the institution's president and board of trustees, as well as the senior administration, faculty leaders, and various academic departments. The strategy for accreditation management should be systematically and attentively planned, harmonized accordingly, budgeted properly, monitored effectively, and truthfully reported. It should be noted that many employment position descriptions in higher education leadership areas are gradually increasing the requirement for successful experience in management of accreditation initiation, renewal, and maintenance. To help address this increasingly needed ability, accreditation guidance and administration should be included in more graduate programs in higher education administration and professional development programs to help guarantee the permanence of this essential method of self-regulation and development for postsecondary institutions in the future.

Academic accreditation has been criticized in recent years, as shown in a variety of reports, by privately funded institutions and publicly elected legislative bodies (Eaton 2003). This unease concentrates on the believed or perhaps real insufficiency of information about institutional performance, ineffective cost restraint, quality control, and inadequate information on the results of student learning. Communication by educational leaders with their internal institutional and outside groups regarding the successful improvements that accreditation efforts and achievement have done for their college or university is very important, along with communicating the changes and improvements that many accreditation associations have enacted in recent years. These improvements in accreditation include the traditional comprehensive reviews, supplementary notice on procedures and verification to evaluate student-learning outcomes of baccalaureate programs, innovative alternatives for focused evaluations, and synchronized accreditation self-evaluations for reduction of expenses and increased review effectiveness. Influential academic leaders should seek to influence the perceptions of stakeholders about the institution and the accreditation procedures positively yet truthfully. This will help ensure that the current system of unique self-regulation that has enabled U.S. colleges and universities to preserve internal approval and overall global superiority.

While we are seeing that accrediting agencies in the United States are addressing the many challenges they face with new options and approaches, the road to the future of accreditation is still not clear. Will it lead toward more regulation and public oversight, remain as it is now, or enable greater self-regulation and innovative reviews for improvement? Higher education has traditionally been run with a high degree of autonomy from outside influences (Berdahl and McConnell 1999). However, the current trend toward increased accountability by the public, legislators, and state budget offices is not likely to abate until society is satisfied that those in higher education have listened to such concerns. More information about, and productivity from, individual professors and institutions at large is sought. In addition, this turning point in the U.S. postsecondary system may influence a global turning point in higher education regulation, program delivery, and demands by the global society. It has been stated that colleges and universities around the world differ according to cultural, political, and economic conditions, yet do share some common traits (Altbach 1999). These traits and even specific academic programs are becoming more common as specialized accreditation in professions such as nursing, business, and engineering, for example, are now assessed globally by accreditors (Lenn and Campos 1997). Therefore, the public in the United States and abroad is seeking more information about institutional quality, cost effectiveness (return on tuition investments), and guidance on selecting institutions to attend.

This author agrees that higher education has a dual role where colleges both scrutinize and serve the society it supports. Therefore, a turn toward increased vocationalization and less self-regulation may negate the dual role. It has been added and opined that "too much autonomy encourages colleges and universities, both public and private, to slight society's needs. Too much accountability produces dependent institutions subservient to society's whims." (McGuinness 2002 in Burke 2005) Balance between independent higher education and accountability is certainly a hopeful part of the future direction so that we do not lose the social and public mission of higher education. Leaders and scholars of higher education have been quite vocal in their conclusions that we have greatly increased the corporatization of administration and research, commercialized athletics, and have sought to economize the system at the expense of purpose and quality (Bok 2003; Kezar 2004). Administrative methods adapted from business are not all inappropriate or non-useful, but when we "attempt to quantify matters within the university that are not quantifiable" (Bok 2003, p. 3), support is given to the internal critics of change which leads to stagnation, bloating, and lethargy that are part of society's ill perceptions (true or not) of academe today. Only through a balanced, thorough, and comprehensive approach to self-regulation, improvement, and peer-review can we earn the trust of the global society.

Accreditation has value, and we in higher education must understand and accept its role in sustaining the quality of postsecondary instruction, research, and service. The current president of the Council of Higher Education Accreditation has stated that accreditation is actually a benefit to academe, in that the accreditation system acts as a buffer against the politicizing of higher education and helps maintain our values (Eaton 2003). This buffer would be a loss if the turning point in the evolution of accreditation leads toward less self-governance, and the future of higher education would most likely be changed toward a different path than the one we are currently headed for. Over the past millennium, higher education has shown strong resilience yet an ability to change when needed. One of the fundamental features of higher education that enables this characteristic is the tenured appointment of faculty and many senior administrators. The tenured appointments tend to protect the status quo and block changes (Duderstadt and Womack 2003). Where many leaders in higher education tend to be cautious and boards are often distracted from strategic issues in favor of personal interests and political agendas, universities have survived largely intact over the centuries in part because of our extraordinary ability to survive, change, and adapt. This author agrees with the belief as stated nearly 25 years ago in a monograph on accreditation (Harcleroad 1980) that the U.S. federal government's tax and spending power will continue to grow, and therefore the more efficient and less costly self-regulated, voluntary associations currently working in the public interest will continue, albeit with some changes in structure, increased public awareness, and accountability.

The present process of accreditation was recently at a crossroads, when the U.S. government representatives directed postsecondary officials to monitor higher education achievement more successfully through self-regulation and prevention of fraud, or else they possibly might receive supplementary government supervision and the removal of the regional system (Glidden 1996). The federal government's escalating influence on the oversight of learning and generally managing higher education will probably cause some grave alarms about what influence this will have on self-governance for colleges and universities. Even though it is probably not in the framework of current legal doctrine, the government may choose to create another oversized control agency, comparable to that many other countries have in a ministry of education (Eaton 2003). Naturally, this supplementary agency budget would add to the financial burden of current and future taxpayers. Contrarily, if the federal government decides to delegate the oversight function for accreditation and approval of postsecondary institutions, that would result in no less than 50 diverse states systems of educational quality evaluation with dissimilar regulations, principles, and objectives. The arrangement of college and university accreditation needs to be supported by the institutions and their leaders using whatever resources they can make available, since it is critical for the individual colleges, universities, and the whole system of higher education that has been and is a very successful system of regulation developed over the past 100 years.

In the first part of this book, the features of the various generations of higher education accreditation that developed over the past century were explained. Now it is important consider what the next generation will incorporate. Questions such as, will there be increased level of accountability along with apparent value because of the enhancements in the mechanism of academic self-governance, arise. Researchers and higher education leaders have declared a variety of rationales and projected several potential outcomes for the future structure of postsecondary accreditation. The changing nature of college student learning, the transformation in teaching styles, such as active learning, using Internet-based online formats, and the requirement for greater public involvement will encourage the system of accreditation to reassess what features are evaluated, how the evaluations are performed, and who is involved in the reviews (Ewell 1998). The next generation of accreditation in higher education will probably include an expansion of the range and methods of the evaluation procedures, while concurrently enabling the accreditation review to require less time for participating institutions. Moreover, there will be an ongoing requirement to create additional capability for performance audits that are founded on college student learning (Eaton 2001). Some writers have contemplated that the principally process-oriented accreditation system that is in use these days is not appropriate as postsecondary institutions become increasingly more involved with adult-centered and nontraditional educational offerings. It has also been proposed that in coming years, a need will arise for very specialized, unconventional accreditation associations, comparable in authority and public respect to the current regional accreditation associations in the United States that may be established to evaluate and accredit nontraditional specialized programs. (Hogg 1993)

It was more than three decades ago that Martin Trow observed and expressed the imminent troubles related to the evolution of higher education from elite to mass to universal access (1973). At present, postsecondary education is challenged by the ongoing "universalization" where larger ratios and raw numbers of students desire to enroll in colleges and universities, combined with increasing managerialism and commercialization, rising globalization, and the omnipresent new instructional technologies that not only influence the instructional content and delivery methods, but also emphasize the increasing difference in ages between faculty and students. The next stage in the evolution of accreditation development will need to address these societal elements that are changing, while concurrently improving the respect of the general public about the quality assurance of higher education. The coming developments in the next generation of accreditation may affect long-established colleges and universities in unforeseen and probably positive ways by gently nudging institutions to focus more on their true student-customer area and mission, and not seek change by attempting to reach service areas beyond their capability and market need. Some people experienced in the field of higher education believe that this often-unnecessary development by individual institutions happens because the desire for prestige on the part of the faculty (or perhaps the senior administrator) often pushes institutions to change program offerings and overall institutional evolution, so that they can be more like the colleges or universities that they truly wish they could be associated with (Calhoun 1999). At times these endeavors are fruitful and appropriate, but occasionally many of these changed institutions are then not able to serve a particular market and therefore miss opportunities in enrollment, grant, and program development that is more directly tied to their original mission, resulting in budget problems and other challenges.

Others hold the view that higher education on the whole has overspecialized with the various specialized academic disciplines. This creates a confusing amalgam for baccalaureate studies in the United States, combined with the latest modularization of courses, and far too many course electives for students to select (Trow 1994). This allows colleges and universities to then change or attempt to make changes that are fundamentally against their original mission and currently served stakeholders, often resulting in less emphasis on effective teaching and more emphasis on faculty research, poor student service, and too much planning in nonacademic programs. The next generation of college accreditation will need to strongly direct that institutions stay with their mission and capabilities, and only make developments and changes in appropriate directions which are supported by the community and are within their means. The next generation of accreditation will also need to fully address the rise of distance learning, which is creating demands to reposition accreditation as a source of information about quality (Eaton 2002). Private magazine guidebook rankings can only have a limited value, especially in a society that increasingly looks to the government for more and more advice, support, and direction. The kind of collegiate information that students, the public, and government are increasingly seeking from accreditation is developing toward more clearly delineated affirmative or negative judgments (summative evaluations) as to whether accreditation assures educational quality. This is a change from the conventional conclusions of accreditation agencies that served more as diagnoses to improve quality (formative evaluations) rather than as summative statements that are of limited use to the institution. However, if the accreditation agencies do provide these yes or no responses that are in demand, the potential college students and others will probably seek out other information sources such as the government, the media, or the private business sector, to acquire the data they believe is needed to make informed decisions about quality. This conundrum for academic accreditation and higher education in general is still evolving, and the societal character and genuineness of higher education are at risk. Accreditation agencies and the many higher education leaders involved in accreditation have a duty to reach beyond the institutions they are closely connected with and seek the betterment of postsecondary generation oversight with responsible, thoughtful, and positive improvements.

When considering the rapidly changing international economy, a significantly higher level of accountability in the business field after a plethora of corporate scandals, ever-increasing tuition costs for college students, and state-level budgetary reductions for higher education, the essential values required to properly direct enhancements in this system should strongly support greater organizational responsibility at colleges and universities, while permitting institutions to have self-guided autonomy that has enabled the great successes which U.S. higher education is known for. Regulatory oversight and government accreditation procedures should strive to "do no harm" to organizational autonomy, allow the great range and diversity of institutional types, reaffirm that academic accountability is a primary goal, and establish a forward-looking system of responsibility (Graham et al 1995). It is also widely held that academic freedom is an essential role and probably one of the most basic issues in institutional consideration for accreditation approval, evaluation processes, and the future development of accreditation evolution (Elman 1994). The significant responsibility of academic freedom in the standards of accreditation approval and higher education organizational planning complements the very critical duty of accreditation criteria in affirming that colleges and universities have academic freedom for teaching, research, and service activities. This mutually beneficial arrangement between colleges and associations is good for educational institutions, faculty positions, student achievement, and the tuition and tax-paying public that they both serve. When considering if there were increased government oversight and manipulation instead of the voluntary self-regulated accreditation system, then this increased direction from the government will probably threaten the basis of academic freedom at institutions and also reduce the overall quality of flourishing higher education institutions today that were established to leverage these ideas.

A large part of the reason why many faculty members and higher education leaders have been reluctant to measure student-learning outcomes and other performance standards is because of the fundamental desire to retain academic freedom. According to the president of the Council of Higher Education Accreditation, accreditation and its procedure has been viewed by many people in the field as a confirmation of specific principles that are essential to the ethos and thoughts about higher education (Eaton 1999). These viewpoints and ideas solidly sustain the tenets of collegiality, peer-evaluation, self-development, and institutional sovereignty. In the coming years, college accreditation will have to tackle these subjects, and also have the ability to require high quality performance of all colleges and universities. Active operational involvement by the institutions in the accreditation self-study evaluation processes, similar to instructional active learning by students that has become commonly used in classrooms, can be highlighted to college community members such as faculty as a mechanism to alter viewpoints and obtain their support for the procedure (Ewell 2001). College accreditation must continue to improve with more recurring and important advice to colleges, and conceivably help institutions into becoming true organizational learning systems, possessing the ability to create and transfer knowledge for ongoing improvement (Dill 2000). The recent proposals for internal audit review systems related to procedures used in other countries will not completely attend to these requirements and the strong demands for external accountability, along with institutional and program breadth. Nevertheless, portions of these audit systems with true answerability that methodically evaluate institutional procedures, often as a supplement to outcomes, could possibly be included in the coming years. Meanwhile, U.S. colleges and universities should understand that accreditation is still very important, and there are many new innovations presented for utilizing and leveraging to assist all institutions in accomplishing the revitalization and public appreciation that they ought to have.

APPENDIX A—Recognized Accrediting Organizations

This chart lists regional, national, and specialized accreditors that are or have been recognized by the Council for Higher Education Accreditation (CHEA) or the U.S. Department of Education (USDE) or both. Organizations identified by (•) are recognized; (–) indicates those not currently recognized. An asterisk (∗) identifies accrediting organizations that were formerly recognized. CHEA recognized organizations must meet CHEA eligibility standards (www.chea.org/recognition/recognition.asp). Accreditors exercise independent judgment about seeking CHEA recognition. For USDE recognition, accreditation from the organization is used by an institution or program to establish eligibility to participate in federal student aid or other federal programs (www.ed.gov/about/offices/list/ope/index.html). Some accreditors cannot be considered for USDE recognition because they do not provide access to federal funds. Other accreditors have chosen not to pursue USDE recognition. Because CHEA affiliation and USDE recognition depend on a range of factors, readers are strongly cautioned against making judgments about the quality of an accrediting organization and its institutions and programs based solely on CHEA or USDE status. Additional inquiry is essential. If you have questions about the CHEA or USDE recognition status of an accreditor, please contact the accrediting organization.

This chart [CHEA, 2006 #874] (used with permission) is updated (on the website) when the CHEA Board of Directors recognizes an accrediting organization and when the United States Secretary of Education recognizes an accrediting organization.

APPENDIX A—Recognized Accrediting Organizations

ACCREDITOR	CHEA Recognized Organization	USDE Recognized Organization
REGIONAL ACCREDITING ORGANIZATIONS		
Middle States Association of Colleges and Schools Middle States Commission on Higher Education	●	●
New England Association of Schools and Colleges Commission on Institutions of Higher Education	●	●
New England Association of Schools and Colleges Commission on Technical and Career Institutions	●	●
North Central Association of Colleges and Schools The Higher Learning Commission	●	●
Northwest Commission on Colleges and Universities	●	●
Southern Association of Colleges and Schools Commission on Colleges	●	●
Western Association of Schools and Colleges Accrediting Commission for Community and Junior Colleges	●	●
Western Association of Schools and Colleges Accrediting Commission for Senior Colleges and Universities	●	●
FAITH-BASED ACCREDITING ORGANIZATIONS		
Association for Biblical Higher Education Commission on Accreditation	●	●
Association of Advanced Rabbinical and Talmudic Schools Accreditation Commission	●	●
Commission on Accrediting of the Association of Theological Schools in the United States and Canada	●	●
Transnational Association of Christian Colleges and Schools Accreditation Commission	●	●
PRIVATE CAREER ACCREDITING ORGANIZATIONS		
Accrediting Bureau of Health Education Schools	—	●
Accrediting Commission of Career Schools and Colleges of Technology	—	●

APPENDIX A—Recognized Accrediting Organizations • 181

Organization	Col1	Col2	Col3
Accrediting Council for Continuing Education and Training	—	•	
Accrediting Council for Independent Colleges and Schools	•	•	
Council on Occupational Education	—	•	
Distance Education and Training Council Accrediting Commission	•	•	
National Accrediting Commission of Cosmetology Arts and Sciences, Inc.	—	•	
SPECIALIZED AND PROFESSIONAL ACCREDITING ORGANIZATIONS			
AACSB International—The Association to Advance Collegiate Schools of Business	•	*	
Accreditation Board for Engineering and Technology, Inc.	•	*	
Accreditation Commission for Acupuncture and Oriental Medicine	—	•	
Accreditation Council for Pharmacy Education	•	•	
Accreditation Review Commission on Education for the Physician Assistant, Inc.	•	—	
Accrediting Council on Education in Journalism and Mass Communications	•	*	
American Academy for Liberal Education	—	•	
American Association for Marriage and Family Therapy Commission on Accreditation for Marriage and Family Therapy Education	•	•	
American Association of Family and Consumer Sciences Council for Accreditation	•	—	
American Association of Nurse Anesthetists Council on Accreditation of Nurse Anesthesia Educational Programs	•	•	
American Bar Association Council of the Section of Legal Education and Admissions to the Bar	—	•	
American Board of Funeral Service Education Committee on Accreditation	•	•	
American College of Nurse-Midwives Division of Accreditation	—	•	

Recognized Accrediting Organizations (continued)

ACCREDITOR	CHEA Recognized Organization	USDE Recognized Organization
American Council for Construction Education Board of Trustees	●	*
American Culinary Federation, Inc. Accrediting Commission	●	*
American Dental Association Commission on Dental Accreditation	—	●
American Dietetic Association Commission on Accreditation for Dietetics Education	●	●
American Library Association Committee on Accreditation	●	*
American Occupational Therapy Association Accreditation Council for Occupational Therapy Education	●	●
American Optometric Association Accreditation Council on Optometric Education	●	●
American Osteopathic Association Commission on Osteopathic College Accreditation	*	●
American Physical Therapy Association Commission on Accreditation in Physical Therapy Education	●	●
American Podiatric Medical Association Council on Podiatric Medical Education	●	●
American Psychological Association Committee on Accreditation	●	●
American Society for Microbiology American College of Microbiology	—	*
American Society of Landscape Architects Landscape Architectural Accreditation Board	●	*
American Speech-Language-Hearing Association Council on Academic Accreditation in Audiology and Speech-Language Pathology	●	●
American Veterinary Medical Association Council on Education	●	●

APPENDIX A—Recognized Accrediting Organizations

Association for Clinical Pastoral Education, Inc., Accreditation Commission	●	—
Association of Collegiate Business Schools and Programs	*	●
Commission on Accreditation of Allied Health Education Programs	*	●
Commission on Accreditation of Healthcare Management Education	●	●
Commission on Collegiate Nursing Education	●	●
Commission on English Language Program Accreditation	●	—
Commission on Massage Therapy Accreditation	●	—
Commission on Opticianry Accreditation	●	—
Council for Accreditation of Counseling and Related Educational Programs	—	●
Council for Interior Design Accreditation	*	●
Council on Aviation Accreditation	—	●
Council on Chiropractic Education Commission on Accreditation	●	●
Council on Education for Public Health	●	—
Council on Naturopathic Medical Education	●	●
Council on Rehabilitation Education Commission on Standards and Accreditation	*	●
Council on Social Work Education Office of Social Work Accreditation and Educational Excellence	*	●
Joint Review Committee on Education Programs in Radiologic Technology	●	●
Joint Review Committee on Educational Programs in Nuclear Medicine Technology	●	●
Liaison Committee on Medical Education	●	—

Recognized Accrediting Organizations (continued)

ACCREDITOR	CHEA Recognized Organization	USDE Recognized Organization
Midwifery Education Accreditation Council	—	●
Montessori Accreditation Council for Teacher Education	—	●
National Accrediting Agency for Clinical Laboratory Sciences	●	●
National Architectural Accrediting Board, Inc.	—	*
National Association of Industrial Technology	●	*
National Association of Nurse Practitioners in Women's Health Council on Accreditation	—	●
National Association of Schools of Art and Design Commission on Accreditation	●	●
National Association of Schools of Dance Commission on Accreditation	●	●
National Association of Schools of Music Commission on Accreditation and Commission on Community/Junior College Accreditation	●	●
National Association of Schools of Public Affairs and Administration Commission on Peer Review and Accreditation	●	—

National Association of Schools of Theatre Commission on Accreditation	●	●
National Council for Accreditation of Teacher Education	●	●
National Environmental Health Science and Protection Accreditation Council	—	*
National League for Nursing Accrediting Commission, Inc.	●	●
National Recreation and Park Association/American Association for Physical Activity and Recreation Council on Accreditation	●	—
Planning Accreditation Board	●	—
Society of American Foresters	●	*
Teacher Education Accreditation Council Accreditation Committee	●	●
United States Conference of Catholic Bishops Commission on Certification and Accreditation	—	*

APPENDIX B–Nonrecognized College Accreditation Agencies

The following accrediting agencies are not recognized by the US Department of Education, the Council for Higher Education Accreditation, and often by various state education departments, UNESCO, or by the education departments or ministries of major countries. In addition, they often generally do not adhere to the Generally Accepted Accrediting Practices (GAAP). The list was complied from various sources (Barrett 2005; Degree.net 2000; Michigan 2005; Wikipedia 2006; www.aju.edu 2006) and primary research. As stated on Degree.net, the agencies vary from a number of genuine education attempts that are striving for legitimacy to many associations created by low-quality postsecondary institutions merely to use the term accreditation for their school. The following list contains many of these associations that have been observed in recent years. However, new accreditation agencies are started quite frequently and therefore this list is not exhaustive. If a reader is seeking to determine if accreditation is recognized, it is important to check with the established governmental agencies such as the US Department of Education, CHEA, respected governmental agencies in other countries, and the UNESCO list (UNESCO 2004).

Academy for the Promotion of International Cultural and Scientific Exchange (APICS). According to Degree.net, this organization has offices in Canada and Switzerland, and the secretary general is Dr. Denis K. Muhilly who was not pleased with statements by Degree.net in the past. The organization had earlier approached quite a few schools to discuss European accreditation and now have a list of accredited members at http://www.apics.com/news.htm. Also known by the German name Akademie fuer Internationale Kultur und Wissenschaftsfoerderung.

Accreditation Association of American College and Universities. An unrecognized agency of which the American University of Hawaii and South West University of the USA claimed to be "chartered members." Both universities claimed that the accreditation association had "also made the requisite applications toward the processes leading to recognitions endorsed by the U.S. Department of Education" (see http://www.swuusa.com/index.htm).

Accrediting Commission for Specialized Colleges, Gas City, Indiana. According to Degree.net, this was established by "Bishop" Gordon Da Costa and his associates (one of whom was Dr. George Reuter, who left to help establish the International Accrediting Commission), from the address of Da Costa's Indiana Northern Graduate School (a dairy farm in Gas City) (Degree.net 2000). In reading the agency's literature, the accrediting policies of the ACSC association appear be very shallow. The single prerequisite for becoming a contender for accreditation was to submit a payment for $110 to the association.

Accrediting Commission International for Schools, Colleges and Theological Seminaries, Beebe, Arkansas. See also: "International Accrediting Commission (IAC) for Schools, Colleges and Theological Seminaries." According to Degree.net, after the IAC was investigated, fined, and prohibited from operation by state regulators in Missouri in 1989, Dr. Reuter retired (Degree.net 2000). It was soon thereafter that the ACI was established in the adjoining state of Arkansas, and distributed an invitation to the former IAC schools proposing certain accreditation by the ACI. The researchers at Degree.net were not aware of any colleges that were denied accreditation and the ACI refused to reveal a list of postsecondary institutions that are accredited. It was found that more than 130 schools claimed accreditation from ACI, the majority of which are ostensibly evangelical Bible schools, with a number of nonreligious schools as well such as Century University, Columbia Southern University, Wisconsin International University, and Western States University (Degree.net 2000).

Accrediting Council for Colleges and Schools. Claimed by Century University as an accreditor, this agency is not locatable and is on two lists of unrecognized agencies (Wikipedia 2006; www.aju.edu 2006).

Accreditation Governing Commission of the United States of America. Located in Washington, D.C., it is listed as an unrecognized agency (Barrett 2005; Michigan 2005; Wikipedia 2006). This accrediting agency association's website states that it is a nonprofit organization with seven member institutions.

Akademie fuer Internationale Kultur und Wissenschaftsfoerderung (also see Association for Promotion of International Cultural and Scientific Exchange, APICS)

Alternative Institution Accrediting Association. Allegedly in Washington, D.C., the association appears on several lists of unrecognized agencies, and is the accreditor of several illegitimate schools.

American Association of Accredited Colleges and Universities. Another fake and untraceable agency, and the claimed accreditor of Ben Franklin Academy according to Degree.net.

American Association of Independent Collegiate Schools of Business. According to onlineuniversityreview.com and Degree.net, an untraceable accreditor claimed by Rushmore University.

American Association of Nontraditional Collegiate Business Schools. Another credible-sounding but untraceable accrediting agency mentioned by Rushmore

University according to Degree.net. Also listed as unrecognized (Degree.net 2000; Wikipedia 2006; www.aju.edu 2006).

American Association of Schools. Located in Temple City, California (http://www.a-aos.org), and on at least two lists of unrecognized agencies (Wikipedia 2006; www.aju.edu 2006).

American Council of Private Colleges and Universities. This agency has a website http://www.acpcu.org, and is identified on several lists of unrecognized accreditation associations (Barrett 2005; Wikipedia 2006; www.aju.edu 2006).

American Education Association for the Accreditation of Schools, Colleges and Universities. This is an accreditation agency that was claimed by the University of America. Currently on several lists of unrecognized agencies and untraceable (Degree.net 2000; www.aju.edu 2006).

American International Commission for Excellence in Higher Education, Inc. This organization cannot be located except on one list of unrecognized agencies (www.aju.edu 2006).

American Psycotherapy Association Board of Psycotherapy Examiners, Katy, Texas. According to Degree.net the association claims to be originally chartered in Florida. It apologizes for, yet does not correct the misspellings in its name.

Arizona Commission of Non-Traditional Private Postsecondary Education. Shown on three lists of unrecognized agencies (Degree.net 2000; Wikipedia 2006; www.aju.edu 2006). According to Degree.net, it was established in the late 1970s by the owners of Southland University, which also declared the university to be a candidate for this supposed accreditation. The agency name was changed after a complaint by the real state agency, the Arizona Commission on Postsecondary Education (also see the Western Council on Non-Traditional Private Post Secondary Education).

Association of Accredited Private Schools. Claimed by several diploma mills and according to several sources, this agency has no telephone listed in Federal Way, Washington. It wrote to many schools in 1997, inviting them to send a $1,000 application fee (Degree.net 2000; www.aju.edu 2006).

Association of Career Training Schools. According to Degree.net there was a slick booklet sent to schools stating, "Have your school accredited with the Association. Why? The Association Seal ... could be worth many $ $ $ to you! It lowers sales resistance, sales costs, [and] improves image."

Association for Distance Learning. On two lists of unrecognized agencies (Barrett 2005; Wikipedia 2006).

Association for Online Academic Excellence. Founded in 1997. Their website (http://216.36.213.212/) states that they are a professional accrediting association, established to uphold and maintain high standards for all levels of online postsecondary education. Identified on several lists as an unrecognized agency (Barrett 2005; Wikipedia 2006; www.aju.edu 2006).

The Association for Online Distance Learning. Accredits Chase University and is on two lists of unrecognized agencies (Wikipedia 2006; www. aju 2006).

Association of World Universities. On a list of unrecognized agencies (www.aju.edu 2006).

Board of Online Universities Accreditation. Established in 1994. Their website (www.boua.org) states they accredit institutions such as Ashwood and Rochville Universities. On three lists of unrecognized accreditors (Barrett 2005; Michigan 2005; Wikipedia 2006).

Central States Council on Distance Education. Listed as an unapproved agency (Barrett 2005; Michigan 2005; Wikipedia 2006).

Commission for the Accreditation of European Non-Traditional Universities. Shown on two lists of unrecognized agencies. The University de la Romande, in England, used to claim accreditation from this agency, which is locatable (Degree.net 2000; www.aju.edu 2006).

Council for the Accreditation of Correspondence Colleges. Several curious schools claimed their accreditation; the agency is supposed to be in Louisiana.

Council for International Education Accreditation. Untraceable. Their previous website (www.ciea.org seems inactive. On two lists of unrecognized agencies (Barrett 2005; Wikipedia 2006).

Council for National Academic Accreditation. The Internet shows institutions such as "Universitas Mons Calpe" in Gibraltar and "GOLF ON TOP The Academy of Golf by M.I.T. Colorado." According to Degree.net, this untraceable agency wrote to schools in 1988 from Cheyenne, Wyoming, offering them the opportunity to be accredited on payment of a fee up to $1,850.

Council on Postsecondary Alternative Accreditation. This is an agency on two lists of unrecognized accreditors (Degree.net 2000; www.aju.edu 2006) and in the literature of Western States University. Western States never responded to requests for the address of their accreditor, according to Degree.net. It appears that the agency name may have been selected to cause confusion with the legitimate United States federal organization previously known as the Council on Postsecondary Accreditation (COPA) described earlier in this book.

Council on Postsecondary Christian Education. No information available, except on Degree.net which states that this association was established by the people who operated LaSalle University and Kent College in Louisiana.

Distance Education Council of America (DECA). On two lists of unrecognized agencies (Degree.net 2000; www.aju.edu 2006) and reported as very similar in name and publications to the legitimate US agency called the Distance Education and Training Council. The DECA agency apparently appeared in Delaware in 1998, with a proposal sent to schools to pay $200 or more for supposed accreditation and $150 more for an "Excellence" rating.

Distance Graduate Accrediting Association. Identified on four lists of unrecognized agencies (Barrett 2005; Michigan 2005; Wikipedia 2006; www.aju.edu 2006).

APPENDIX B–Nonrecognized College Accreditation Agencies • 191

Distance Learning Council of Europe. Identified on three lists of unrecognized agencies (Barrett 2005; Michigan 2005; Wikipedia 2006).

European Committee for Home and Online Education. Their website (http://www.echoe.org) claims to represent privately owned and nongovernmental European organizations and they are identified as an unaccredited agency (Wikipedia 2006).

European Council for Distance and Open Learning. Identified on three lists of unrecognized agencies (Barrett 2005; Michigan 2005; Wikipedia 2006).

Higher Education Accreditation Commission. Identified on three lists of unrecognized agencies (Barrett 2005; Michigan 2005; Wikipedia 2006).

Higher Education Services Association Identified on three lists of unrecognized agencies (Barrett 2005; Michigan 2005; Wikipedia 2006).

Inter-Collegiate Joint Committee on Academic Standards. Identified on three lists of unrecognized agencies (Barrett 2005; Michigan 2005; Wikipedia 2006).

Integra Accreditation Association. Identified on two lists of unrecognized agencies (Wikipedia 2006; www.aju.edu 2006).

InterAmerican Association of Postsecondary Colleges and Schools. According to Degree.net, this is the agency from which the Universitas Sancti Martin claims accreditation. The agency is not locatable.

International Academic Accrediting Commission. Identified on two lists of unrecognized agencies (Wikipedia 2006; www.aju.edu 2006).

International Accreditation Association. This agency is listed on CredentialWatch.org and other sites (Barrett 2005; Michigan 2005; Wikipedia 2006) as an unrecognized agency. The literature of the University of North America claims that they are accredited by this association and Degree.net reports that no address is provided, nor could one be located.

International Accreditation Commission for Post Secondary Education Institutions. The University of the United States and Nasson University claim accreditation from this agency which, and Degree.net reports, according to the schools' apparently shared website, has "not sought specific recognition from any single nation."

International Accreditation Agency for Online Universities. Identified on three lists of unrecognized agencies (Barrett 2005; Michigan 2005; Wikipedia 2006).

International Accreditation for Universities, Colleges and Institutes. Identified on three lists of unrecognized agencies (Barrett 2005; Michigan 2005; Wikipedia 2006).

International Accrediting Association. An apparently common name. Degree.net states that the address in Modesto, California, is the same as that of the Universal Life Church, an organization that awards doctoral degrees of all kinds, including the PhD, to anyone making a "donation" of $5 to $100.

International Accrediting Association for Colleges and Universities. Accredits institutions such as Barron University and Huntington Pacific University. Identified on three lists of unrecognized agencies (Barrett 2005; Michigan 2005; Wikipedia 2006).

International Accrediting Commission for Postsecondary Educational Institutions. Very similar in name to the International Accrediting Commission for Postsecondary Institutions. Reported by several lists (Wikipedia 2006; www.aju.edu 2006) and by Degree.net as an unrecognized agency from which the Adam Smith University has previously claimed accreditation.

International Accrediting Commission for Schools, Colleges and Theological Seminaries, Holden, Missouri. A widely used agency reported on several lists of unrecognized agencies. It is reported to accredit more than 150 schools, many of them Bible schools. According to Degree.net, in 1989 the attorney general of Missouri performed a clever "sting" operation, in which a fictitious school, the "East Missouri Business College," was created and rented a one-room office in St. Louis. This supposed school issued a typewritten catalog with interesting college leaders such as "Peelsburi Doughboy" and "Wonarmmd Mann." In addition, the Three Stooges were all on the faculty and the college's official marine biology text was *The Little Golden Book of Fishes*. Nevertheless, Dr. George Reuter, director of the accrediting agency (the IACSCTS) visited the college, accepted their payment, and subsequently accredited the school. Shortly thereafter the IAC was forbidden from conducting business and served with a substantial fine. It was after these events that Dr. Reuter decided to retire, and the almost identical "Accrediting Commission International" appeared at once in Arkansas, and proposed instant accreditation to all IACSCTS members (Degree.net 2000).

International Association of Non-Traditional Schools. This agency is on at least two lists of unrecognized agencies. It is the claimed accreditor of several British degree mills, and allegedly located in England, according to Degree.net.

International Association of Universities and Schools. Accredits organizations such as Trinity International, InterAmerican University, and others. Identified on three lists of unrecognized agencies (Barrett 2005; Michigan 2005; Wikipedia 2006).

International Commission for the Accreditation of Colleges and Universities. This supposed accrediting agency was created in Gaithersburg, Maryland, by a collegiate diploma mill called the United States University of America (now defunct) largely for the function of accrediting themselves (Degree.net 2000).

International Commission for Excellence in Higher Education, Inc. According to Degree.net, Monticello University's Internet publication, this agency "was formed by the Board of Trustees of Monticello University to ensure the highest possible standards of academic excellence in curriculum design and operational policies for member universities. Monticello University is the first distance learning institution to be accredited by this agency." Other member institutions could be found but the agency cannot be located.

International Commission for Higher Education. Founded in 2002 (www.icfhe.org), this organization is identified on three lists of unrecognized agencies (Barrett 2005; Michigan 2005; Wikipedia 2006).

International Commission of Open Post Secondary Education. West Coast University claims this accreditation, and the agency is identified on three lists of unrecognized agencies (Barrett 2005; Michigan 2005; Wikipedia 2006).

International University Accrediting Association. Identified on three lists of unrecognized agencies (Barrett 2005; Michigan 2005; Wikipedia 2006).

Middle States Accrediting Board. Identified on several lists of unrecognized accreditation agencies (Degree.net 2000; Wikipedia 2006; www.aju.edu 2006). Basically it is nonexistent according to Degree.net, and created by Thomas University and other degree mills for the purpose of self-accreditation. Apparently the name was selected to create confusion with the legitimate US regional association called the Middle States Association of Colleges and Schools, in Philadelphia.

Midwestern States Accreditation Agency. Identified on two lists of unrecognized agencies (Wikipedia 2006; www.aju.edu 2006).

National Academy of Higher Education. Their website (http://www.nahighered.org/accreditation.htm) states that the Association of Distance Learning Programs (ADLP), a division of NAHE, is an accrediting association founded to provide a consistent measurement of the acceptability of private schools (K-12), adult high schools, vocational and technical schools, private colleges, and postsecondary education. This agency is on two lists of unrecognized agencies (Barrett 2005; Wikipedia 2006).

National Accreditation Association. On several lists of unrecognized accreditation agencies (Degree.net 2000; Wikipedia 2006; www.aju.edu 2006). According to Degree.net it was established in Riverdale, Maryland, by Dr. Glenn Larsen. Dr. Larsen's doctorate is from a diploma mill called the Sussex College of Technology, and his associate is Dr. Clarence Franklin, former president and chancellor of American International University (another diploma mill). It is known that in a mailing to presidents of unaccredited schools, the NAA offered full accreditation by mail, with no on-site inspection required (Degree.net 2000).

National Association for Private Post-Secondary Education, Washington, D.C. Identified on several lists of unrecognized accreditors (Degree.net 2000; Wikipedia 2006; www.aju.edu 2006) and mentioned in the literature of Kennedy-Western University in 1990. According to Degree.net, the agency states they are not an accrediting agency and instead are a private association of schools, even though Kennedy-Western claimed accreditation from this association.

National Association of Alternative Schools and Colleges. Identified on several lists, (Degree.net 2000; Wikipedia 2006; www.aju.edu 2006). Western States University claimed in their literature that they had been accredited by this organization, which has not been located (Degree.net 2000).

National Association of Open Campus Colleges. Southwestern University of Arizona and Utah (which closed after its proprietor was sent to prison as a result of the Federal Bureau of Investigation's diploma mill investigations) claimed accreditation from this agency (Degree.net 2000). The address in Springfield, Missouri, was the same as that of Disciples of Truth, an organization that has in the past operated a chain of diploma mills. This agency is untraceable and is on several unrecognized agency lists (Degree.net 2000; Wikipedia 2006; www.aju.edu 2006).

The National Association for Private Nontraditional Schools and Colleges. (Formerly the National Association for Schools and Colleges) It was an earnest attempt to create an accrediting agency expressly related to alternative schools and programs (Degree.net 2000). Degree.net states that this agency was established in Grand Junction, Colorado, in the 1970s by a group of educators connected to Western Colorado University, a nontraditional school that has since gone out of business. Currently identified on several lists of unrecognized agencies (Degree.net 2000; Wikipedia 2006; www.aju.edu 2006).

National Commission on Higher Education. Identified on three lists of unrecognized agencies (Barrett 2005; Michigan 2005; Wikipedia 2006).

National Council for the Accreditation of Private Universities and Schools of Law. An unrecognized agency from which Monticello University has claimed accreditation, according to Degree.net.

National Council of Schools and Colleges. Identified on three lists of unrecognized agencies (Degree.net 2000; Wikipedia 2006; www.aju.edu 2006). According to Degree.net, International University, formerly of New Orleans, later of Pasadena, California, and now out of existence, claimed accreditation by this agency. Attempts were made to obtain information from owners of the school about the agency, but no information was provided.

National Distance Learning Accreditation Council. Established in 1995. Their website is http://www.n-d-l-a-c.com/pages/1/index.htm. Identified on three lists of unrecognized agencies (Barrett 2005; Michigan 2005; Wikipedia 2006).

Non-Traditional Course Accreditation Body. Identified on two lists of unrecognized agencies (Barrett 2005; Wikipedia 2006).

Pacific Association of Schools and Colleges – According to Degree.net, this agency was established in 1993 and is (or was) operated by a man who was previously a senior official in the California Department of Education (and holds a doctorate from an unaccredited school). The Federal Register Online (Register 1995) shows an application for approval and Degree.net states that PASC appears (or appeared) to be a serious attempt to create an accreditor that would be designed for nontraditional schools. However, the agency is identified on three lists of unrecognized agencies (Degree.net 2000; Wikipedia 2006; www.aju.edu 2006).

United Congress of Colleges. Located in Ireland. It is on three lists of unrecognized agencies (Barrett 2005; Michigan 2005; Wikipedia 2006).

Universal Council for Online Education Accreditation. Founded in 1995 (www.ucoea.org), it is on three lists of unrecognized agencies (Barrett 2005; Michigan 2005; Wikipedia 2006).

US-DETC. Not to be confused with the legitimate DETC located in Washington, D.C. On three lists of unrecognized agencies (Barrett 2005; Michigan 2005; Wikipedia 2006).

West European Accrediting Society. According to Degree.net, this agency was established from a mail-forwarding service in Liederbach, Germany by the owners of a chain of diploma mills such as Loyola, Roosevelt, Lafayette, Southern California, and Oliver Cromwell universities, for the objective of accrediting these supposed schools. Also identified on three lists of unrecognized agencies (Degree.net 2000; Wikipedia 2006; www.aju.edu 2006).

Western Association of Private Alternative Schools. One of several accrediting agencies claimed in the literature of Western States University, according to Ddegree.net. Also listed on other unrecognized agency lists (Degree.net 2000; Wikipedia 2006; www.aju.edu 2006); it is untraceable.

Western Association of Schools and Colleges. This is the name of the legitimate US regional accreditation agency for California and other western states. However, it is also the name used by the previously mentioned owners of diploma mills such as Loyola and Roosevelt from a Los Angeles address to provide supposed accreditation (Degree.net 2000; Wikipedia 2006; www.aju.edu 2006).

Western Council on Non-Traditional Private Post Secondary Education. According to Degree.net, this is an accrediting agency created by the originators of Southland University, probably for accrediting these supposed colleges and others (also see the Arizona Commission on this listing).

World Association of Universities and Colleges (WAUC). This agency is located in Beverly Hills, California, and has a website (http://www.waucglobalaccreditation.org). Degree.net reports that in February 1995, the national investigative magazine, *Spy*, ran quite an unflattering article on WAUC (Degree.net 2000). The agency website states they have been in existence since 1993, and currently claims to have been working toward recognition by the US Department of Education for the past five years. The website also states that WAUC is recognized by the Association Internationale Des Educateurs Pour La Paix Mondiale that they claim is affiliated with UNESCO and UNICEF. In addition, WAUC states that their accredited schools are recognized by various governments and agencies around the globe.

Worldwide Accrediting Commission. Identified on lists of unrecognized agencies (Degree.net 2000; www.aju.edu 2006). Degree.net reports that this agency operated from a mail-forwarding service in Cannes, France, for the function of accrediting the phony Loyola University (Paris), Lafayette University, and other U.S.-run college diploma mills.

Unrecognized Accreditation Agencies for Health-Related Education and Training

Stephen Barrett, M.D. has tabulated a list on www.credentialwatch.org of nonrecognized accreditation agencies that includes a number of health-related organizations not recognized by the U.S. Department of Education. These include:

American Association of Drugless Practitioners Commission on Accreditation

American Association of International Medical Graduates (also listed in Michigan 2005)

American Naturopathic Medical Certification and Accreditation Board

Commission on Medical Denturitry Accreditation

Council on Medical Denturitry Education

Examining Board of Natural Medicine Practitioners (Barrett 2005)

Other unrecognized health-related agencies:
Central States Consortium of Colleges & Schools (Wikipedia 2006)

Faith-Based and Bible School Accreditation Agencies

There are a number of recognized accreditation agencies of religious schools listed by the CHEA in Appendix A, including associations for various evangelical Christian schools and rabbinical seminaries. Degree.net declares that religious schools often state that they did not pursue accreditation because there are no relevant accreditation agencies. However, this is not entirely correct (Degree.net 2000). Actually there are legitimate agencies as well as a good number of unrecognized accreditation agencies. Because many religion-oriented schools openly proclaim that the diplomas awarded are not truly academic, accreditation for these programs has a different connation. A number of these accreditation agencies are indeed authentic. However, their accreditation approval has no academic significance in a comparable sense to traditional postsecondary education. Some of the accreditation associations in this area are seemingly focused on doctrinal accuracy and occasionally may have other reasons for existing. According to Degree.net, selected faith-based accreditation agencies include:

Accreditation Association of Christian Colleges and Seminaries, Morgantown, KY

Accrediting Association of Christian Colleges and Seminaries, Sarasota, FL

AF Sep (not known what this means, but Beta International University claims it is the name of their accrediting association), address unknown

American Association of Accredited Colleges and Universities

American Association of Theological Institutions

American Educational Accrediting Association of Christian Schools

American Federation of Christian Colleges and Schools

Association of Christian Schools and Colleges

Association of Fundamental Institutes of Religious Education

APPENDIX B–Nonrecognized College Accreditation Agencies

International Accrediting Commission, Kenosha, WI

International Accrediting Association of Church Colleges

National Educational Accrediting Association, Columbus, OH

Southeast Accrediting Association of Christian Schools, Colleges and Seminaries, Milton, FL

World-Wide Accreditation Commission of Christian Educational Institutions (Barrett 2005; Degree.net 2000; Michigan 2005; Wikipedia 2006; www.aju.edu 2006)

Other faith-based unrecognized agencies:

American Accrediting Association of Theological Institutions (Wikipedia 2006)

Association of Christian Colleges and Theological Schools. Based in Virginia (Barrett 2005; Michigan 2005)

Association of Christian Colleges and Theological Schools, Louisiana. (Wikipedia 2006)

Association of Christian Schools International. Offices located in the United States, Canada, and several other countries. Their website (http://www.acsi.org/) lists current membership at over 5,440 member schools in 105 countries worldwide. Identified as an unrecognized accreditor (Wikipedia 2006).

Global Accreditation Commission. Located in San Diego, California, and described at http://www.fivefold-ministry.com. Listed as unrecognized (Wikipedia 2006; www.aju.edu 2006).

International Interfaith Accreditation Association. Their websites (www.iiaaweb.org) lists two seminaries and four colleges and universities.

Southern Accrediting Association of Bible Institutes and Colleges. Accredits Kingsway Christian College and is on three lists of unrecognized agencies (Barrett 2005; Michigan 2005; Wikipedia 2006).

APPENDIX C—Accreditation Eligibility Requirements

The essential eligibility requirements that must be met for consideration of Candidacy for Accreditation status, and a sample is shown below for a selected regional accrediting association in the United States, the Higher Learning Commission of the North Central Association of Colleges and Schools (NCA 2003).

The Criteria for Accreditation are organized under five major headings. Each Criterion has three elements: Criterion Statement, Core Components, and Examples of Evidence. The Criteria Statements define necessary attributes of an organization accredited by the Commission. An organization must be judged to have met each of the Criteria to merit accreditation. An organization addresses each Core Component as it presents reasonable and representative evidence of meeting a Criterion. The Examples of Evidence illustrate the types of evidence an organization might present in addressing a core component.

The Criteria are intentionally general so that accreditation decisions focus on the particulars of each organization, rather than on trying to make it fit a preestablished mold. The widely different purposes and scopes of colleges and universities demand criteria that are broad enough to encompass diversity and support innovation, but clear enough to ensure acceptable quality.

The Criteria Statements and Core Components are presented here. Visit the Commission's website (http://www.ncahigherlearningcommission.org) to view the Examples of Evidence.

Criterion 1: Mission and Integrity. The organization operates with integrity to ensure the fulfillment of its mission through structures and processes that involve the board, administration, faculty, staff, and students.

   a. The organization's mission documents are clear and articulate publicly the organization's commitments.
   b. In its mission documents, the organization recognizes the diversity of its learners, other constituencies, and the greater society it serves.
   c. Understanding of and support for the mission pervade the organization.

d. The organization's governance and administrative structures promote effective leadership and support collaborative processes that enable the organization to fulfill its mission.
   e. The organization upholds and protects its integrity.

Criterion 2: Preparing for the Future. The organization's allocation of resources and its processes for evaluation and planning demonstrate its capacity to fulfill its mission, improve the quality of its education, and respond to future challenges and opportunities.

   a. The organization realistically prepares for a future shaped by multiple societal and economic trends.
   b. The organization's resource base supports its educational programs and its plans for maintaining and strengthening their quality in the future.
   c. The organization's ongoing evaluation and assessment processes provide reliable evidence of institutional effectiveness that clearly informs strategies for continuous improvement.
   d. All levels of planning align with the organization's mission, thereby enhancing its capacity to fulfill that mission.

Criterion 3: Student Learning and Effective Teaching. The organization provides evidence of student learning and teaching effectiveness that demonstrates its fulfilling of its educational mission.

   a. The organization's goals for student-learning outcomes are clearly stated for each educational program and make effective assessment possible.
   b. The organization values and supports effective teaching.
   c. The organization creates effective learning environments.
   d. The organization's learning resources support student learning and effective teaching.

Criterion 4: Acquisition, Discovery, and Application of Knowledge. The organization promotes a life of learning for its faculty, administration, staff, and students by fostering and supporting inquiry, creativity, practice, and social responsibility in ways consistent with its mission.

   a. The organization demonstrates, through the actions of its board, administrators, students, faculty, and staff, that it values a life of learning.
   b. The organization demonstrates that acquisition of a breadth of knowledge and skills and the exercise of intellectual inquiry are integral to its educational programs.
   c. The organization assesses the usefulness of its curricula to students who will live and work in a global, diverse, and technological society.
   d. The organization provides support to ensure that faculty, students, and staff acquire, discover, and apply knowledge responsibly.

Criterion 5: Engagement and Service. As called for by its mission, the organization identifies its constituencies and serves them in ways both value.

   a. The organization learns from the constituencies it serves and analyzes its capacity to serve their needs and expectations.
   b. The organization has the capacity and the commitment to engage with its identified constituencies and communities.
   c. The organization demonstrates its responsiveness to those constituencies that depend on it for service.
   d. Internal and external constituencies value the services the organization provides.

(NCA 2003, used with permission)

APPENDIX D—Institutional Newsletter Content Prior to Accreditation Visit

Volume	Areas Covered
Volume 1, Issue 1	The North Central Association: Who Are They? The Five Criteria for NCA Accreditation: A Summary Forms of Affiliation The Evaluation Process
Volume 1, Issue 2	NCA Accreditation: Frequently Asked Questions The Relationships between GIRs and the Criteria
Volume 1, Issue 3	1997–98 College Self-Study Report: Do You Know? NCA Criterion One: Patterns of Evidence The College's Request for Institutional Change: Additional Criteria What Are Your Questions?
Volume 1, Issue 4	Guidelines for Distance Education: What the NCA Expects Criterion Two: Patterns of Evidence Distance Education Definition Curriculum and Instruction Evaluation and Assessment Library and Learning Resources Student Services Facilities and Finances Question of the Week: What is the Commission Accrediting?
Volume 1, Issue 5	Assessment of Student Academic Achievement and Institutional Effectiveness Criterion Three: Patterns of Evidence The 1997–98 Self-Study Report: Section that Focuses on Criterion Three

	Evidence of Student Learning
	Evidence of Institutional Effectiveness
Volume 1, Issue 6	Meet the NCA Team: A Brief Profile
	Criterion Four: Patterns of Evidence
	Criterion Five: Patterns of Evidence

Source: McCallin 1999, p. 304. ©Commission on Colleges and Universities of the Northwest Association of Schools and of Colleges and Universities. Used with permission.

APPENDIX E—Accreditation in the United States

Accreditation in the United States (USDE 2006, p. 900), Subpart B—The Criteria for Recognition.

Basic eligibility requirements

602.10   Link to Federal programs.

The agency must demonstrate that—

(a) If the agency accredits institutions of higher education, its accreditation is a required element in enabling at least one of those institutions to establish eligibility to participate in HEA programs; or
(b) If the agency accredits institutions of higher education or higher education programs, or both, its accreditation is a required element in enabling at least one of those entities to establish eligibility to participate in non-HEA Federal programs. (Authority: 20 U.S.C. 1099b)

602.11   Geographic scope of accrediting activities.

The agency must demonstrate that its accrediting activities cover—

(a) A State, if the agency is part of a State government;
(b) A region of the United States that includes at least three States that are reasonably close to one another; or
(c) The United States.
(Authority: 20 U.S.C. 1099b)

602.12 Accrediting experience.

(a) An agency seeking initial recognition must demonstrate that it has—
   (1) Granted accreditation or pre-accreditation—
      (i) To one or more institutions if it is requesting recognition as an institutional accrediting agency and to one or more programs if it is requesting recognition as a programmatic accrediting agency;
      (ii) That it covers the range of the specific degrees, certificates, institutions, and programs for which it seeks recognition;
      (iii) Is in the geographic area for which it seeks recognition; and
   (2) Conducted accrediting activities, including deciding whether to grant or deny accreditation or pre-accreditation, for at least two years prior to seeking recognition.
(b) A recognized agency seeking an expansion of its scope of recognition must demonstrate that it has granted accreditation or pre-accreditation covering the range of the specific degrees, certificates, institutions, and programs for which it seeks the expansion of scope.
(Authority: 20 U.S.C. 1099b)

602.13 Acceptance of the agency by others.

The agency must demonstrate that its standards, policies, procedures, and decisions to grant or deny accreditation are widely accepted in the United States by—

(a) Educators and educational institutions; and
(b) Licensing bodies, practitioners, and employers in the professional or vocational fields for which the educational institutions or programs within the agency's jurisdiction prepare their students.
(Authority: 20 U.S.C. 1099b)

Organizational and administrative requirements

602.14 Purpose and organization.

(a) The Secretary recognizes only the following four categories of agencies:

The Secretary recognizes	that
(1) An accrediting agency	(i) Has a voluntary membership of institutions of higher education; (ii) Has as a principal purpose—the accrediting of institutions of higher education and that accreditation is a required element in enabling those institutions to participate in HEA programs; and (iii) Satisfies the separate and independent requirements in paragraph (b) of this section.

The Secretary recognizes	that
(2) An accrediting agency	(i) Has a voluntary membership; and (ii) Has as its principal purpose the accrediting of higher education programs, or higher education programs and institutions of higher education, and that accreditation is a required element in enabling those entities to participate in non-HEA Federal programs.
(3) An accrediting agency	for purposes of determining eligibility for Title IV, HEA programs— (i) Either has a voluntary membership of individuals participating in a profession or has as its principal purpose the accrediting of programs within institutions that are accredited by a nationally recognized accrediting agency; and (ii) Either satisfies the separate and independent requirements in paragraph (b) of this section or obtains a waiver of those requirements under paragraphs (d) and (e) of this section.
(4) A State agency	(i) Has as a principal purpose—the accrediting of institutions of higher education, higher education programs, or both; and (ii) The Secretary listed as a nationally recognized accrediting agency on or before October 1, 1991 and has recognized continuously since that date.

(b) For purposes of this section, the term "separate and independent" means that—
   (1) The members of the agency's decision-making body—who decide the accreditation or pre-accreditation status of institutions or programs, establish the agency's accreditation policies, or both—are not elected or selected by the board or chief executive officer of any related, associated, or affiliated trade association or membership organization;
   (2) At least one member of the agency's decision-making body is a representative of the public, and at least one-seventh of that body consists of representatives of the public;

(3) The agency has established and implemented guidelines for each member of the decision-making body to avoid conflicts of interest in making decisions;
(4) The agency's dues are paid separately from any dues paid to any related, associated, or affiliated trade association or membership organization; and
(5) The agency develops and determines its own budget, with no review by or consultation with any other entity or organization.
(c) The Secretary considers that any joint use of personnel, services, equipment, or facilities by an agency and a related, associated, or affiliated trade association or membership organization does not violate the separate and independent requirements in paragraph (b) of this section if—
(1) The agency pays the fair market value for its proportionate share of the joint use; and
(2) The joint use does not compromise the independence and confidentiality of the accreditation process.
(d) For purposes of paragraph (a)(3) of this section, the secretary may waive the "separate and independent" requirements in paragraph (b) of this section if the agency demonstrates that—
(1) The Secretary listed the agency as a nationally recognized agency on or before October 1, 1991, and has recognized it continuously since that date;
(2) The related, associated, or affiliated trade association or membership organization plays no role in making or ratifying either the accrediting or policy decisions of the agency;
(3) The agency has sufficient budgetary and administrative autonomy to carry out its accrediting functions independently; and
(4) The agency provides to the related, associated, or affiliated trade association or membership organization only information it makes available to the public.
(e) An agency seeking a waiver of the "separate and independent" requirements under paragraph (d) of this section must apply for the waiver each time the agency seeks recognition or continued recognition.
(Authority: 20 U.S.C. 1099b)

602.15 Administrative and fiscal responsibilities.

The agency must have the administrative and fiscal capability to carry out its accreditation activities in light of its requested scope of recognition. The agency meets this requirement if the agency demonstrates that—
(a) The agency has—
(1) Adequate administrative staff and financial resources to carry out its accrediting responsibilities;
(2) Competent and knowledgeable individuals, qualified by education and experience in their own right and trained by the agency on its standards, policies, and procedures, to conduct its on-site evaluations, establish its policies, and make its accrediting and pre-accrediting decisions;

(3) Academic and administrative personnel on its evaluation, policy, and decision-making bodies, if the agency accredits institutions;
(4) Educators and practitioners on its evaluation, policy, and decision-making bodies, if the agency accredits programs or single-purpose institutions that prepare students for a specific profession;
(5) Representatives of the public on all decision-making bodies; and
(6) Clear and effective controls against conflicts of interest, or the appearance of conflicts of interest, by the agency's
   (i) Board members;
   (ii) Commissioners;
   (iii) Evaluation team members;
   (iv) Consultants;
   (v) Administrative staff; and
   (vi) Other agency representatives; and
(b) The agency maintains complete and accurate records of—
   (1) Its last two full accreditation or pre-accreditation reviews of each institution or program, including on-site evaluation team reports, the institution's or program's responses to on-site reports, periodic review reports, any reports of special reviews conducted by the agency between regular reviews, and a copy of the institution's most recent self-study; and
   (2) All decisions regarding the accreditation and pre-accreditation of any institution or program, including all correspondence that is significantly related to those decisions.
(Approved by the Office of Management and Budget under control number 1845-0003)
(Authority: 20 U.S.C. 1099b)

Required standards and their application

602.16 *Accreditation and pre-accreditation standards.*

(a) The agency must demonstrate that it has standards for accreditation, and pre-accreditation, if offered, that are sufficiently rigorous to ensure that the agency is a reliable authority regarding the quality of the education or training provided by the institutions or programs it accredits. The agency meets this requirement if—
   (1) The agency's accreditation standards effectively address the quality of the institution or program in the following areas:
      (i) Success with respect to student achievement in relation to the institution's mission, including, as appropriate, consideration of course completion, state licensing examination, and job placement rates.
      (ii) Curricula.
      (iii) Faculty.
      (iv) Facilities, equipment, and supplies.
      (v) Fiscal and administrative capacity as appropriate to the specified scale of operations.

(vi) Student support services.
(vii) Recruiting and admissions practices, academic calendars, catalogs, publications, grading, and advertising.
(viii) Measures of program length and the objectives of the degrees or credentials offered.
(ix) Record of student complaints received by, or available to, the agency.
(x) Record of compliance with the institution's program responsibilities under Title IV of the act, based on the most recent student loan default rate data provided by the secretary, the results of financial or compliance audits, program reviews, and any other information that the secretary may provide to the agency; and

(2) The agency's pre-accreditation standards, if offered, are appropriately related to the agency's accreditation standards and do not permit the institution or program to hold pre-accreditation status for more than five years.

(b) If the agency only accredits programs and does not serve as an institutional accrediting agency for any of those programs, its accreditation standards must address the areas in paragraph (a)(1) of this section in terms of the type and level of the program rather than in terms of the institution.

(c) If none of the institutions an agency accredits participates in any Title IV, HEA program, or if the agency only accredits programs within institutions that are accredited by a nationally recognized institutional accrediting agency, the agency is not required to have the accreditation standards described in paragraphs (a)(1)(viii) and (a)(1)(x) of this section.

(d) An agency that has established and applies the standards in paragraph (a) of this section may establish any additional accreditation standards it deems appropriate.

(Approved by the Office of Management and Budget under control number 1845-0003)

(Authority: 20 U.S.C. 1099b)

602.17 *Application of standards in reaching an accrediting decision.*

The agency must have effective mechanisms for evaluating an institution's or program's compliance with the agency's standards before reaching a decision to accredit or pre-accredit the institution or program. The agency meets this requirement if the agency demonstrates that it—

(a) Evaluates whether an institution or program—
(1) Maintains clearly specified educational objectives that are consistent with its mission and appropriate in light of the degrees or certificates awarded;
(2) Is successful in achieving its stated objectives; and
(3) Maintains degree and certificate requirements that at least conform to commonly accepted standards;

(b) Requires the institution or program to prepare, following guidance provided by the agency, an in-depth self-study that includes the assessment of educational quality and the institution's or program's continuing efforts to improve educational quality;

(c) Conducts at least one on-site review of the institution or program during which it obtains sufficient information to determine if the institution or program complies with the agency's standards;

(d) Allows the institution or program the opportunity to respond in writing to the report of the on-site review;

(e) Conducts its own analysis of the self-study and supporting documentation furnished by the institution or program, the report of the on-site review, the institution's or program's response to the report, and any other appropriate information from other sources to determine whether the institution or program complies with the agency's standards; and

(f) Provides the institution or program with a detailed written report that assesses
   (1) The institution's or program's compliance with the agency's standards, including areas needing improvement; and
   (2) The institution's or program's performance with respect to student achievement.

(Authority: 20 U.S.C. 1099b)

602.18 *Ensuring consistency in decision making.*

The agency must consistently apply and enforce its standards to ensure that the education or training offered by an institution or program, including any offered through distance education, is of sufficient quality to achieve its stated objective for the duration of any accreditation or pre-accreditation period granted by the agency. The agency meets this requirement if the agency—

(a) Has effective controls against the inconsistent application of the agency's standards;

(b) Bases decisions regarding accreditation and pre-accreditation on the agency's published standards; and

(c) Has a reasonable basis for determining that the information the agency relies on for making accrediting decisions is accurate.

(Authority: 20 U.S.C. 1099b)

602.19 *Monitoring and reevaluation of accredited institutions and programs.*

(a) The agency must reevaluate, at regularly established intervals, the institutions or programs it has accredited or pre-accredited.

(b) The agency must monitor institutions or programs throughout their accreditation or pre-accreditation period to ensure that they remain in compliance with the agency's standards. This includes conducting special evaluations or site visits, as necessary.

(Authority: 20 U.S.C. 1099b)

602.20   Enforcement of standards.

(a) If the agency's review of an institution or program under any standard indicates that the institution or program is not in compliance with that standard, the agency must—
  (1) Immediately initiate adverse action against the institution or program; or
  (2) Require the institution or program to take appropriate action to bring itself into compliance with the agency's standards within a time period that must not exceed—
    (i) Twelve months, if the program, or the longest program offered by the institution, is less than one year in length;
    (ii) Eighteen months, if the program, or the longest program offered by the institution, is at least one year, but less than two years, in length; or
    (iii) Two years, if the program, or the longest program offered by the institution, is at least two years in length.
(b) If the institution or program does not bring itself into compliance within the specified period, the agency must take immediate adverse action unless the agency, for good cause, extends the period for achieving compliance.
(Authority: 20 U.S.C. 1099b)

602.21   Review of standards.

(a) The agency must maintain a systematic program of review that demonstrates that its standards are adequate to evaluate the quality of the education or training provided by the institutions and programs it accredits and relevant to the educational or training needs of students.
(b) The agency determines the specific procedures it follows in evaluating its standards, but the agency must ensure that its program of review—
  (1) Is comprehensive;
  (2) Occurs at regular, yet reasonable, intervals or on an ongoing basis;
  (3) Examines each of the agency's standards and the standards as a whole; and
  (4) Involves all of the agency's relevant constituencies in the review and affords them a meaningful opportunity to provide input into the review.
(c) If the agency determines, at any point during its systematic program of review, that it needs to make changes to its standards, the agency must initiate action within 12 months to make the changes and must complete that action within a reasonable period of time. Before finalizing any changes to its standards, the agency must—
  (1) Provide notice to all of the agency's relevant constituencies, and other parties who have made their interest known to the agency, of the changes the agency proposes to make;
  (2) Give the constituencies and other interested parties adequate opportunity to comment on the proposed changes; and
  (3) Take into account any comments on the proposed changes submitted timely by the relevant constituencies and by other interested parties.
(Authority: 20 U.S.C. 1099b)

Required operating policies and procedures

602.22 Substantive change.

(a) If the agency accredits institutions, it must maintain adequate substantive change policies that ensure that any substantive change to the educational mission, program, or programs of an institution after the agency has accredited or pre-accredited the institution, does not adversely affect the capacity of the institution to continue to meet the agency's standards. The agency meets this requirement if—
   (1) The agency requires the institution to obtain the agency's approval of the substantive change before the agency includes the change in the scope of accreditation or pre-accreditation it previously granted to the institution; and
   (2) The agency's definition of substantive change includes at least the following types of change:
      (i) Any change in the established mission or objectives of the institution.
      (ii) Any change in the legal status, form of control, or ownership of the institution.
      (iii) The addition of courses or programs that represent a significant departure, in either content or method of delivery, from those that were offered when the agency last evaluated the institution.
      (iv) The addition of courses or programs at a degree or credential level above that which is included in the institution's current accreditation or pre-accreditation.
      (v) A change from clock hours to credit hours.
      (vi) A substantial increase in the number of clock or credit hours awarded for successful completion of a program.
      (vii) The establishment of an additional location geographically apart from the main campus at which the institution offers at least 50 percent of an educational program.
(b) The agency may determine the procedures it uses to grant prior approval of the substantive change. Except as provided in paragraph (c) of this section, these may, but need not, require a visit by the agency.
(c) If the agency's accreditation of an institution enables the institution to seek eligibility to participate in Title IV, HEA programs, the agency's procedures for the approval of an additional location described in paragraph (a)(2)(vii) of this section must determine if the institution has the fiscal and administrative capacity to operate the additional location. In addition, the agency's procedures must include—
   (1) A visit, within six months, to each additional location the institution establishes, if the institution—
      (i) Has a total of three or fewer additional locations;
      (ii) Has not demonstrated, to the agency's satisfaction, that it has a proven record of effective educational oversight of additional locations; or

(iii) Has been placed on warning, probation, or show cause by the agency or is subject to some limitation by the agency on its accreditation or pre-accreditation status;
(2) An effective mechanism for conducting, at reasonable intervals, visits to additional locations of institutions that operate more than three additional locations; and
(3) An effective mechanism, which may, at the agency's discretion, include visits to additional locations, for ensuring that accredited and pre-accredited institutions that experience rapid growth in the number of additional locations maintain educational quality.

(d) The purpose of the visits described in paragraph (c) of this section is to verify that the additional location has the personnel, facilities, and resources it claimed to have in its application to the agency for approval of the additional location.

(Authority: 20 U.S.C. 1099b)

602.23 Operating procedures all agencies must have.

(a) The agency must maintain and make available to the public, upon request, written materials describing—
  (1) Each type of accreditation and pre-accreditation it grants;
  (2) The procedures that institutions or programs must follow in applying for accreditation or pre-accreditation;
  (3) The standards and procedures it uses to determine whether to grant, reaffirm, reinstate, restrict, deny, revoke, terminate, or take any other action related to each type of accreditation and pre-accreditation that the agency grants;
  (4) The institutions and programs that the agency currently accredits or pre-accredits and, for each institution and program, the year the agency will next review or reconsider it for accreditation or pre-accreditation; and
  (5) The names, academic and professional qualifications, and relevant employment and organizational affiliations of—
    (i) The members of the agency's policy and decision-making bodies; and
    (ii) The agency's principal administrative staff.

(b) In providing public notice that an institution or program subject to its jurisdiction is being considered for accreditation or pre-accreditation, the agency must provide an opportunity for third-party comment concerning the institution's or program's qualifications for accreditation or pre-accreditation. At the agency's discretion, third-party comment may be received either in writing or at a public hearing, or both.

(c) The accrediting agency must—
  (1) Review in a timely, fair, and equitable manner any complaint it receives against an accredited institution or program that is related to the agency's standards or procedures;
  (2) Take follow-up action, as necessary, including enforcement action, if necessary, based on the results of its review; and

(3) Review in a timely, fair, and equitable manner, and apply unbiased judgment to, any complaints against itself and take follow-up action, as appropriate, based on the results of its review.
(d) If an institution or program elects to make a public disclosure of its accreditation or pre-accreditation status, the agency must ensure that the institution or program discloses that status accurately, including the specific academic or instructional programs covered by that status and the name, address, and telephone number of the agency.
(e) The accrediting agency must provide for the public correction of incorrect or misleading information an accredited or pre-accredited institution or program releases about—
   (1) The accreditation or pre-accreditation status of the institution or program;
   (2) The contents of reports of on-site reviews; and
   (3) The agency's accrediting or pre-accrediting actions with respect to the institution or program.
(f) The agency may establish any additional operating procedures it deems appropriate. At the agency's discretion, these may include unannounced inspections.
(Approved by the Office of Management and Budget under control number 1845-0003)
(Authority: 20 U.S.C. 1099b)

602.24 Additional procedures certain institutional accreditors must have.

If the agency is an institutional accrediting agency and its accreditation or pre-accreditation enables those institutions to obtain eligibility to participate in Title IV, HEA programs, the agency must demonstrate that it has established and uses all of the following procedures:

(a) Branch campus.
   (1) The agency must require the institution to notify the agency if it plans to establish a branch campus and to submit a business plan for the branch campus that describes—
      (i) The educational program to be offered at the branch campus;
      (ii) The projected revenues and expenditures and cash flow at the branch campus; and
      (iii) The operation, management, and physical resources at the branch campus.
   (2) The agency may extend accreditation to the branch campus only after it evaluates the business plan and takes whatever other actions it deems necessary to determine that the branch campus has sufficient educational, financial, operational, management, and physical resources to meet the agency's standards.
   (3) The agency must undertake a site visit to the branch campus as soon as practicable, but no later than six months after the establishment of that campus.

(b) Change in ownership.

The agency must undertake a site visit to an institution that has undergone a change of ownership that resulted in a change of control as soon as practicable, but no later than six months after the change of ownership.

(c) Teach-out agreements.
   (1) The agency must require an institution it accredits or pre-accredits that enters into a teach-out agreement with another institution to submit that teach-out agreement to the agency for approval.
   (2) The agency may approve the teach-out agreement only if the agreement is between institutions that are accredited or pre-accredited by a nationally recognized accrediting agency, is consistent with applicable standards and regulations, and provides for the equitable treatment of students by ensuring that—
      (i) The teach-out institution has the necessary experience, resources, and support services to provide an educational program that is of acceptable quality and reasonably similar in content, structure, and scheduling to that provided by the closed institution; and
      (ii) The teach-out institution demonstrates that it can provide students access to the program and services without requiring them to move or travel substantial distances.
   (3) If an institution the agency accredits or pre-accredits closes, the agency must work with the Department and the appropriate State agency, to the extent feasible, to ensure that students are given reasonable opportunities to complete their education without additional charge.
   (Approved by the Office of Management and Budget under control number 1845-0003)
   (Authority: 20 U.S.C. 1099b)

602.25 Due process.

The agency must demonstrate that the procedures it uses throughout the accrediting process satisfy due process. The agency meets this requirement if the agency does the following:

(a) The agency uses procedures that afford an institution or program a reasonable period of time to comply with the agency's requests for information and documents.
(b) The agency notifies the institution or program in writing of any adverse accrediting action or an action to place the institution or program on probation or show cause. The notice describes the basis for the action.
(c) The agency permits the institution or program the opportunity to appeal an adverse action and the right to be represented by counsel during that appeal. If the agency allows institutions or programs the right to appeal other types of actions, the agency has the discretion to limit the appeal to a written appeal.

(d) The agency notifies the institution or program in writing of the result of its appeal and the basis for that result.
(Authority: 20 U.S.C. 1099b)

602.26 Notification of accrediting decisions.

The agency must demonstrate that it has established and follows written procedures requiring it to provide written notice of its accrediting decisions to the Secretary, the appropriate State licensing or authorizing agency, the appropriate accrediting agencies, and the public. The agency meets this requirement if the agency, following its written procedures—

(a) Provides written notice of the following types of decisions to the Secretary, the appropriate State licensing or authorizing agency, the appropriate accrediting agencies, and the public no later than 30 days after it makes the decision:
  (1) A decision to award initial accreditation or pre-accreditation to an institution or program.
  (2) A decision to renew an institution's or program's accreditation or pre-accreditation;
(b) Provides written notice of the following types of decisions to the secretary, the appropriate State licensing or authorizing agency, and the appropriate accrediting agencies at the same time it notifies the institution or program of the decision, but no later than 30 days after it reaches the decision:
  (1) A final decision to place an institution or program on probation or an equivalent status.
  (2) A final decision to deny, withdraw, suspend, revoke, or terminate the accreditation or pre-accreditation of an institution or program;
(c) Provides written notice to the public of the decisions listed in paragraphs (b)(1) and (b)(2) of this section within 24 hours of its notice to the institution or program;
(d) For any decision listed in paragraph (b)(2) of this section, makes available to the Secretary, the appropriate State licensing or authorizing agency, and the public upon request, no later than 60 days after the decision, a brief statement summarizing the reasons for the agency's decision and the comments, if any, that the affected institution or program may wish to make with regard to that decision; and
(e) Notifies the Secretary, the appropriate State licensing or authorizing agency, the appropriate accrediting agencies, and, upon request, the public if an accredited or pre-accredited institution or program—
  (1) Decides to withdraw voluntarily from accreditation or pre-accreditation, within 30 days of receiving notification from the institution or program that it is withdrawing voluntarily from accreditation or pre-accreditation; or
  (2) Lets its accreditation or pre-accreditation lapse, within 30 days of the date on which accreditation or pre-accreditation lapses.

(Approved by the Office of Management and Budget under control number 1845-0003)
(Authority: 20 U.S.C. 1099b)

602.27 Other information an agency must provide the department.

The agency must submit to the Department—

(a) A copy of any annual report it prepares;
(b) A copy, updated annually, of its ulectory of accredited and pre-accredited institutions and programs;
(c) A summary of the agency's major accrediting activities during the previous year (an annual data summary), if requested by the Secretary to carry out the Secretary's responsibilities related to this part;
(d) Any proposed change in the agency's policies, procedures, or accreditation or pre-accreditation standards that might alter its—
  (1) Scope of recognition; or
  (2) Compliance with the criteria for recognition;
(e) The name of any institution or program it accredits that the agency has reason to believe is failing to meet its Title IV, HEA program responsibilities or is engaged in fraud or abuse, along with the agency's reasons for concern about the institution or program; and
(f) If the Secretary requests, information that may bear upon an accredited or preaccredited institution's compliance with its Title IV, HEA program responsibilities, including the eligibility of the institution or program to participate in Title IV, HEA programs. The Secretary may ask for this information to assist the Department in resolving problems with the institution's participation in the Title IV, HEA programs.

(Approved by the Office of Management and Budget under control number 1845-0003)
(Authority: 20 U.S.C. 1099b)

602.28 Regard for decisions of States and other accrediting agencies.

(a) If the agency is an institutional accrediting agency, it may not accredit or pre-accredit institutions that lack legal authorization under applicable State law to provide a program of education beyond the secondary level.
(b) Except as provided in paragraph (c) of this section, the agency may not grant initial or renewed accreditation or pre-accreditation to an institution, or a program offered by an institution, if the agency knows, or has reasonable cause to know, that the institution is the subject of—
  (1) A pending or final action brought by a State agency to suspend, revoke, withdraw, or terminate the institution's legal authority to provide postsecondary education in the State;
  (2) A decision by a recognized agency to deny accreditation or pre-accreditation;

(3) A pending or final action brought by a recognized accrediting agency to suspend, revoke, withdraw, or terminate the institution's accreditation or pre-accreditation; or

(4) Probation or an equivalent status imposed by a recognized agency.

(c) The agency may grant accreditation or pre-accreditation to an institution or program described in paragraph (b) of this section only if it provides to the secretary, within 30 days of its action, a thorough and reasonable explanation, consistent with its standards, why the action of the other body does not preclude the agency's grant of accreditation or pre-accreditation.

(d) If the agency learns that an institution it accredits or pre-accredits, or an institution that offers a program it accredits or pre-accredits, is the subject of an adverse action by another recognized accrediting agency or has been placed on probation or an equivalent status by another recognized agency, the agency must promptly review its accreditation or pre-accreditation of the institution or program to determine if it should also take adverse action or place the institution or program on probation or show cause.

(e) The agency must, upon request, share with other appropriate recognized accrediting agencies and recognized State approval agencies information about the accreditation or pre-accreditation status of an institution or program and any adverse actions it has taken against an accredited or pre-accredited institution or program.

(Approved by the Office of Management and Budget under control number 1845-0003)

(Authority: 20 U.S.C. 1099b)

APPENDIX F—Sample Guidelines for a Regional Accreditation Self-Study Report

(ACCJC 2005) (used with permission)

Development of the Comprehensive Self Study Report

Each institution affiliated with the Accrediting Commission for Community and Junior Colleges accepts the obligation to undergo periodic evaluation through self study and professional peer review. The heart of this obligation is the conducting of a rigorous self study during which an institution appraises itself in terms of the Commission Standards in accord with its stated purposes. A Comprehensive Self Study is required every six years following initial accreditation. The Commission's expectation on periodic review, found in the Accreditation Reference Handbook, governs conditions under which an institution is periodically evaluated.

The Self Study Manual, intended for use with the Guide to Evaluating Institutions (Standards adopted June 2002), provides a reference for the conduct of the comprehensive self study. The Guide to Evaluating Institutions is a document meant to provoke thoughtful consideration about whether the institution meets the Accreditation Standards at a deeper level than mere compliance. The Guide contains the Standards followed by questions to use in institutional evaluation. These questions provide an interpretation of the standards and how they might be applied to an institution, creating the context for a holistic, systemic assessment of the institution.

Self study is part of a three-part process of accrediting an institution. This process includes an institutional self appraisal, an on-site visit by a team of peers, and a review and a decision on the accredited status of the college by the Commission. The institutional self appraisal results in a Report that is an analysis of the ongoing and systematic activities and achievements of an institution. The aim of self appraisal is to assess how well an institution meets Accreditation Standards, Eligibility Requirements, and policies of the Commission and to stimulate improvement of educational quality and institutional performance. The ultimate goal of accreditation is to help an institution improve attainment of its own mission-improving student learning and student achievement.

Self appraisal requires a conscious and self-reflective analysis of strengths and weaknesses and an examination of every aspect of institutional function against Commission Standards. Continuous dialogue among members of the college community—a dialogue that is consistently central to institutional processes and which serves to provide the college community with the means for arriving at a comprehensive institutional perspective—can be especially valuable as the institution engages in self study preparatory to writing a report. Broad involvement in the both the institutional self appraisal and preparation of the Self Study Report enhances the credibility and usefulness of the self study report.

Participation in the Self Study

Included in the self study document submitted to the Commission is a certification page (Appendix A) bearing the signatures of institutional leaders and attesting to broad participation in self appraisal and preparation of the Self Study Report. The certification page reflects the belief that the Self Study Report accurately portrays the nature and substance of the institution. Since the inclusion of all constituencies of the college ensures that the self study does not reflect the exclusive view of any one group, the visiting evaluation team will seek to confirm that all campus constituents have participated in the work of the self study.

Students

Although obtaining broad and representative participation from students is often difficult, student leaders are typically enthusiastic participants on the steering committee. Every effort should be made to enlist student participation.

Faculty

All faculty have a major role to play in the self study process. The faculty perspective on the integrity, quality, and effectiveness of the institution is an integral part of the self study document. Adjunct faculty should be included in the process to the extent possible.

Staff

Support staff must be included in the self study. Employees in all quarters of the institution are knowledgeable about the college and can offer a perspective on how the college is functioning in terms of its stated purposes and Commission Standards. Recognizing the contributions of this constituency is important, as is including them as active participants in the process.

Administrators

Administrators must share in the work of the self study, collaborating with faculty, staff, and students in the search for evidence that the institution meets Commission Standards. The perspective of administrators is an important part of a self study.

Participation in the Self Study

Trustees

Governing Board participation can take a variety of forms. Progress reports on the self study are a way to secure Board participation. Note that at the conclusion of the self study, the Board must certify both participation in the process and the Self Study Report.

Others

The institution may elect to include others in the self study such as members of foundation boards, program advisory committee members, or others. Care should be taken in these selections to avoid the perception of conflicts of interest.

The Commission Standards

The four Commission Standards work together in an integrated way and several themes thread throughout them. These themes can provide guidance and structure to self-reflective dialogue and evaluation of institutional effectiveness as the institution prepares its self study. The themes include:

- institutional commitments to providing high-quality education congruent with institutional mission, to focusing on student learning, and to periodic reflection on the mission statement;
- evaluation, planning, and improvement in an ongoing and systematic cycle that includes evaluation, goal setting, resource distribution, implementation, and reevaluation;
- student learning outcomes as the conscious and robust demonstration of the effectiveness of institutional efforts to produce and support student learning by developing student learning outcomes at the course, program, certificate, and degree level;
- organization as demonstrated in having adequate staff, resources, and organizational structure (communication and decision-making structures) to identify and make public learning outcomes, to evaluate the effectiveness of programs in producing those outcomes, and to make improvements.
- dialogue as a means to ongoing participation in institutional self-reflection based on reliable information about the college's programs and services and evidence on how well the institution is meeting student needs;
- institutional integrity demonstrated through concern with honesty, truthfulness, and the manner in which the institution represents itself to all stakeholders, internal and external.

(For a more complete discussion of these themes, see Guide to Evaluating Institutions) (ACCJC, 2005b).

Preparation for a self study and a Self Study Report under these integrated standards requires that attention be given to weaving these themes with responses given to specific a standard and its subparts. Those charged with the structuring of the process for doing the self study should be mindful of the importance of organizing working committees to address the standards in a coherent way that leads to holistic assessment of institutional quality.

Calendar for Preparation of the Self Study

Since the date for the evaluation visit is often set more than a year in advance, a realistic and detailed timetable for the organization and completion of the self study report should be developed. In most instances, at least a year and a half should be allowed and, for many colleges, there is an advantage to beginning the activities a full four semesters before the scheduled visitation.

A convenient and effective method for establishing a calendar is to work back from the date set for the team visit. In this way, target dates can be set for the completion of activities and the amount of time necessary for meeting goals can be better estimated. Note that the completed self study must be in the hands of the Commission and the team members six weeks before the scheduled visit date.

Several target dates should be kept in mind while planning the calendar. Time needs to be allowed for evidence gathering and interpretation, review of drafts along the way, final editing and rewriting, board of trustees review, and publication. The work of the editor(s) should produce a coherent document that reflects perspectives developed through the process of dialogue.

Resources for the Self Study

Since evaluation and planning are continuous activities complementing and supporting the self study, the Accrediting Commission encourages institutions to integrate the self study with ongoing evaluation and planning, making the six-year self study a culminating activity rather than an activity undertaken only in the last few months before a team visit. Accreditation standards require ongoing program review. These data and analyses are a good source for self appraisal.

A primary goal of the self study should be to provide evidence of institutional effectiveness and compliance with Commission standards. This goal requires that the study include data on students and their learning outcomes. All research and other activities reporting student achievement and learning outcomes done by the institution (formal and informal) since the last visit should be reported. Information on good evidence can be found in the Commission's Guide to Evaluating Institutions.

Another source of data on outcomes can be found in public institutions and institutions that are part of a system because they generate considerable information in the form of reports to system, state, or federal authorities. Vocational, specially funded, or specially accredited programs, for example, sometimes have reporting requirements that generate valuable data on outcomes.

Because institutions must generate and utilize information in ways and forms that are most useful to them in meeting their institutional purposes, the Commission is more interested in how colleges integrate information into their planning process than in the compilation of unanalyzed reports. Creating new reports specifically for the self study is not necessary.

Most institutions routinely and systematically analyze local and regional demographic data. City and county planning offices, associations of regional governments, state government, U.S. census, local school districts, public utilities, business and trade organizations, and other planning interests commonly produce much pertinent data.

In an effort to provide a forum in which individuals and institutions may profit from the experiences of others, the Commission presents self study workshops each year that are designed to assist institutions as they begin to develop their self studies. This forum offers an opportunity for a good deal of interaction with the Commission. The individuals charged with directing the self study should attend this workshop.

The Accreditation Liaison Officer as a Resource

The Accreditation Liaison Officer (ALO) is the individual appointed by the College to serve as the contact between the campus and the Commission. The ALO assumes responsibility for:

The Self Study

- Attending the self study workshop.
- Facilitating the development of the Self Study Report.
- Facilitating distribution of the Self Study Report.
- Facilitating the team visit.
- Facilitating follow-up with the Commission.

Ongoing activities

- Staying informed about Commission policies, procedures, and activities
- Promoting a campus culture that is concerned about accreditation.
- Promoting a campus culture that focuses on student-learning outcomes.
- Acting as an archivist for the institution's accreditation documents.
- Facilitating preparation of the annual reports and other reports to the Commission.
- Facilitating reports on Substantive Change.

Format and Content for the Comprehensive Self Study Report

1. Cover Sheet
   The cover sheet should include the name and address of the institution, a notation that the self study is in support of an application for candidacy, accreditation, or reaffirmation, and date submitted (see Appendix B).

2. Certification Page
   The certification page should include the names of the institutional leaders and attesting to a broad participation in the Report preparation (see Appendix A.)

3. Table of Contents
4. An Introduction
   a. A history of the institution, including a concise and factual description of the institution since the last comprehensive visit.
   b. Demographic information, including summary data on the area served, enrollment figures, and student and staff diversity, including trends and available projections should be provided.
   c. A discussion of the results of the last comprehensive visit, including evidence of what the institution has done regarding the previous team's recommendations. Each recommendation should be addressed separately.
   d. Longitudinal student achievement data, including information on course completion transfer rates, number of degrees and certificates awarded, job placement, licensure, persistence rates, retention rates, graduation rates, basic skills completion, success after transfer, etc.
   e. The Commission recognizes institutions are in varying stages of developing and assessing student learning outcomes at the course, program, and degree level. The college should describe evidence gathered to-date, how it is being used, and what plans exist for continued expansion of this effort.
   f. Information regarding off-campus sites and centers as well as distance-learning efforts should be included. Teams are charged with assuring the commission of quality of all programs.
   g. Information regarding an external independent audit and information demonstrating integrity in the use of federal grant monies.
5. Abstract of the Report
   The Abstract should be a summative assessment of how well the institution is meeting the standards as a whole. It should be based on the themes that pervade the standards: institutional commitments; evaluation, planning, and improvement; student-learning outcomes; organization; dialogue; and institutional integrity.
6. Organization for the Self Study
   In narrative or chart form, this section should show the organization established to conduct the self study. Committees, their chairs and members, timetable, and the person(s) responsible for the overall direction of the self study should be included.
7. Organization of the Institution
   Organization charts for the institution and for each major function should be included. Names of individuals holding each position should be provided. Institutions in multi-college districts/systems must specify whether primary responsibility for all or parts of a specific function is at the college or district level. This organizational "map" is important in evaluating the quality of performance of that function and establishing accountability for doing so. Those who are responsible should be involved in reporting about the function and be held accountable for its improvement. As a result, close cooperation between and among the institutions and the district/system office is expected as a part

8. Certification of continued compliance with Eligibility Requirements
   The institution should summarize the review conducted to verify that it continues to meet eligibility requirements. Specific guidance for this requirement can be found in Appendix C. These pages include the requirements themselves as well as what documents are needed to verify continued eligibility. The college should develop a statement for each of the 21 criteria. The President and the Chair of the Governing Board must sign a statement certifying compliance.

9. Responses to Recommendations from the Most Recent Comprehensive Evaluation
   The self study report must include a section that concisely indicates what the institution has done regarding recommendations made in the last comprehensive team report. Recommendations represent the observations and analyses of a visiting team at the time of the visit and should be considered in light of the Commission's standards and the institution's mission. Evaluation team members will review the responses to previous recommendations.

10. Institutional Self-Evaluation Using Commission Standards
    The primary portion of a self study report reviews institutional performance using the accreditation standards and their themes. The following three elements should guide how the self study report is written.

Descriptive Summary

This narrative should spring from institutional dialogue and should be focused on evidence the college has amassed in support of assertions about what it does to meet Commission standards. The underlying question regards what the institution has learned/knows about what it does.

Self Evaluation

The institution is expected to analyze and systematically evaluate what it has learned/knows about itself in terms of the standards. The basic questions have to do with whether or not and to what degree institutional evidence demonstrates that the institution meets the standards and how the institution has reached this conclusion. This analysis should result in actionable conclusions about institutional effectiveness and capacity, informing decisions for what needs to be done to improve.

Planning Agenda

As an institution describes and evaluates its programs and services with reference to each standard, it identifies areas in need of change. This activity yields a planning agenda — a vehicle for institutional improvement. As the institution assesses itself, it should forecast progress it plans to make. The planning agenda should include the following elements:

   a. Statements of the plans, activities, and processes (as opposed to tasks) the institution expects to implement as a statement of what the institution thinks it will do.
   b. Discussion of the ways the areas identified in need of improvement will be or have been incorporated into the ongoing, systematic evaluation and planning processes of the institution.
   c. Discussion of how the outcomes of these plans, activities, and processes are expected to improve student learning and foster institutional improvement in general.

Note:

The standards reference specific Commission policies. The self study report should address how the college is in compliance with these policies. A list of these policies will be found in Appendix D. Text of the policies can be found in the Accreditation Reference Handbook.

11. A List of the Evidence Available in the Team Room
    Evidence available to the visiting team should include primary sources and reports on which the Self Study Report is based. When evidence is cited in the Report, it should be indexed by standard for easy reference by team members. The Guide to Evaluating Institutions contains many suggestions regarding evidence. The visiting evaluation team will rely heavily on the evidence provided to it in the Team Room and elsewhere.

Tips for the Preparation of the Self-Study Report

Following are some suggestions for conducting the self study and preparing the Self Study Report.

About Participation

The Commission's emphasis on inclusive institutional dialogue as a continuous process sets the tone for participation in self study and the development of a self study report. Basically, the college is expected to provide evidence of broad participation and a commitment to making a concerted effort to providing the opportunity for all voices to be heard in the self study effort.

The Steering Committee

This committee should assume responsibility for overall planning and supervision of the self study report. The membership of the committee can be drawn from existing committee structures of the college currently being used as a means for conducting institutional dialogue. The committee should be given time to assume this responsibility and the clerical support needed to complete its work. The committee should have easy access to evidence and research.

Writing and Editing the Self Study Report

Given the structure and integrated nature of the commission standards and the themes, there are several ways that institutions could configure the work of their committees. One way would be to organize committees utilizing the themes. The six committees would write to the substandards that fit an assigned theme. Membership should include individuals from all constituencies of the college. This arrangement would make holistic weaving of themes and standards part of the process of writing the self-study report and would yield a product that addresses both the structure of the standards and the manner in which they are integrated.

Another way to organize would be to create four committees, one for each standard. In this model, subcommittees would address the standards, using the themes as the overarching structure. Once again, the weaving of standards and themes would provide a holistic approach to thinking and writing about the institution, producing a Self Study Report that uses the integrated standards and themes as its underpinnings.

Tips for Preparing the Self Study Report

Whatever model the institution chooses to employ, sharing information across committees is very important and serves to diminish the likelihood of a Self Study Report that is lacking integration and coherency. Circulating drafts among all constituencies of the college through use of technology is a way to encourage multiple voices as well as greater integration of information and evidence.

It is advantageous to select an editor for the Self Study Report early so that he/she can participate throughout process.

Submission of the Self Study Report

After certification by college constituencies and review by the governing board, four copies of the Self Study Report, four catalogs, and four class schedules should be sent to the Accrediting Commission office. The Commission also requires one electronic version of the Self Study Report. A copy of the report, a catalog, and a schedule should be sent to each member of the evaluation team. Distribution of the report should occur at least six weeks prior to the scheduled evaluation visit. Copies of the report should be made available to members of the college community and to the governing board.

REFERENCES

AAM (1989). *Accreditation: A Handbook for the Visiting Committee.* Washington, D.C.: American Association of Museums.
ACCJC (2005a). *Self Study Manual.* Novato, CA: Accrediting Commission for Community and Junior Colleges, Western Association of Schools and Colleges.
——— (2005b). "Guide to Evaluating Institutions." http://www.accjc.org/ACCJC_Publications.htm.
Allen, I. Elaine, and Jeff Seaman (2005). *Growing by Degrees: Online Education in the United States.* The Sloan Consortium. www.sloan-c.org.
Alstete, Jeffrey W. (1996a). *Benchmarking in Higher Education: Adapting Best Practices to Improve Quality.* San Francisco: Jossey-Bass.
——— (1996b). "Competitive Benchmarking of Non-Credit Program Administration." presentation at the 58th Annual Meeting of the Association for Continuing Higher Education. Palm Desert, CA.
——— (1997). "The Correlates of Administrative Decentralization." *Journal of Education for Business* 73 (1): 21–28.
——— (2001). "Alternative Uses of Electronic Learning Systems for Enhancing Team Performance," *Team Performance Management: An International Journal* 7 (3–4).
——— (2004). *Accreditation Matters: Achieving Academic Recognition and Renewal.* San Francisco: Jossey-Bass.
Altbach, Philip G. (1999). "Patterns in Higher Education Development." In *American Higher Education in the Twenty-First Century: Social, Political, and Economic Challenges,* ed. Philip G. Altbach, Robert O. Berdahl, and Patricia J. Gumport. Baltimore: Johns Hopkins University Press.
Amaral, Alberto M. S. C. (1998). "The US Accreditation System and the CRE's Quality Audits—A Comparative Study." *Quality Assurance in Education* 6 (4): 184–196.
Andersen, Charles J. (1987). *Survey of Accreditation Issues.* Washington, D.C.: American Council on Education.
AQIP (2002). "Understanding AQIP." The Higher Learning Commission of the North Central Association of Schools and Colleges. www.aqip.org/doc/UnderstandingAQIP.pdf
Armstrong, Robert L. (1994). "Outcomes Accreditation: A Response to Innovation and Institutional Change." *NCA Quarterly* 68 (3): 379–385.
Askins, Donna L. (2003). "Utilizing Online Course Software to Prepare a Higher Learning Commission Self-Study." In *108th Annual Meeting of the North Central Association,* ed. Susan E. Van Kollenberg. Chicago, IL: The Higher Learning Commission.

Astin, Alexander W. (2004). "To Use Graduation Rates to Measure Excellent, You Have To Do Your Homework" *Chronicle of Higher Education* 51 (9): 9.

Atkins, Thomas, and Gene Nelson (2001). "Plagiarism and the Internet: Turning the Tables." *English Journal* 90 (4): 101–104.

Bal, Jay, and John Gundy (1999). "Virtual Teaming in the Automotive Supply Chain." *Team Performance Management: An International Journal* 5 (6).

Barker, Thomas S., and Howard W. Smith, Jr. (1998). "Integrating Accreditation into Strategic Planning." *Community College Journal of Research and Practice* 22 (8): 741–750.

Baron, Julie, and Steven M. Crooks (2005). "Academic Integrity in Web Based Distance Education." *TechTrends Linking Research and Practice to Improve Learning* 29 (2): 40–45.

Barr, Robert B., and John Tagg (1995). "From Teaching to Learning—A New Paradigm for Undergraduate Education." *Change* 27 (6): 13–25.

Barrett, Stephen (2005). "Nonrecognized Accreditation Agencies." Credential Watch. http://www.credentialwatch.org/non/agencies.shtml

Bartelt, Carol, and Carol Mishler (1996). "Using Quality Tools to Get Started with Your Self-Study." In *A Collection of Papers on Self-Study and Institutional Improvement*. Chicago, IL: Commission on Institutions of Higher Education of the North Central Association of Colleges and Schools.

Bartlett, Thomas (2004). "Fighting Fakery: Colloquy Live." *Chronicle of Higher Education*. www.chronicle.com.

Bartlett, Thomas, and Scott Smallwood (2004). "Degrees of Suspicion." *Chronicle of Higher Education* 50 (42): A(9).

Beanland, D. (2001). "The Australian Approach toward Quality Enhancement." *Society for Research into Higher Education* 46 (1): 4–5.

Bell, Daniel (1973). *The Coming of the Post-Industrial Society: A Venture in Social Forecasting*. New York: Basic Books.

Bender, Tisha (2003). *Discussion-Based Online Teaching to Enhance Student Learning: Theory, Practice and Assessment*. San Francisco: Jossey-Bass.

Benjamin, Ernst (1994). "From Accreditation to Regulation: The Decline of Academic Autonomy in Higher Education," *ACADEME* (July–August): 34–36.

Bennion, Donald, George Liepa, and Patrick Melia (2002). "The Selection, Care, and Feeding of the Steering Committee: A Key to Successful Self-Study." In *A Collection of Papers on Self-Study and Institutional Improvement*. Chicago, IL: The Higher Learning Commission of the North Central Association of Colleges and Schools.

Benson, P. George (2004). "The Evolution of Business Education in the U.S." *Decision Line* 35 (1): 17–20. http://www.decisionsciences.org/DecisionLine/Vol35/35_1/35_1dean.pdf.

Berdahl, Robert O., and T. R. McConnell (1999). "Autonomy and Accountability: Who Controls Academe?" In *American Higher Education in the Twenty-First Century: Social, Political, and Economic Challenges*, ed. Philip G. Altbach, Robert O. Berdahl, and Patricia J. Gumport. Baltimore: Johns Hopkins University Press.

Bers, Trudy, and Sarah B. Lindquist (2002). "Institutional Research: An Antidote to Accreditation Anxiety." In *A Collection of Papers on Self-Study and Institutional Improvement*. Chicago, IL: The Higher Learning Commission of the North Central Association of Colleges and Schools.

Birch, Garnet E. (1979). "State Higher Education Agency Responsibility for the Evaluation and Accreditation of Public Four-Year Institutions of Higher Education." PhD diss., University of Arizona.

Birnbaum, Robert (1988). *How Colleges Work: The Cybernetics of Academic Organization and Leadership*. San Francisco: Jossey-Bass.

——— (2000). *Management Fads in Higher Education: Where They Come From, What They Do, Why They Fail.* San Francisco: Jossey-Bass.

Bishop, Jane, Jerrilyn Brewer, Dennis Ladwig, and Lee Rasch (2000). "Do We Get Them? Do We Keep Them? Do They Learn? Applying Quality Principles to Higher Education." In *A Collection of Papers on Self-Study and Institutional Improvement.* Chicago, IL: Commission on Institutions of Higher Education of the North Central Association of Colleges and Schools.

Bishop, Jane A., Dennis Ladwig, and Michael Lanser (2001). "Preparing Your Organization for AQIP." In *A Collection of Papers on Self-Study and Institutional Improvement.* Chicago, IL: The Higher Learning Commission of the North Central Association of Schools and Colleges.

Blauch, Lloyd E., ed. (1959). *Accreditation in Higher Education.* New York: Greenwood Press.

Blazey, Mark (1998). "Insights into Organizational Self-Assessments." *Quality Progress* 31 (10): 47–52.

Blazey, Mark, Karen S. Davison, and John P. Evans (2003). *Insights to Performance Excellence in Education, 2003: An Inside Look at the 2003 Baldrige Award Criteria for Education.* Milwaukee, WI: ASQ Press.

Bloland, Harland G. (2001). *Creating the Council for Higher Education Accreditation.* Phoenix, AZ: American Council on Education and the Oryx Press.

Boeree, C. George (1998). "Abraham Maslow." Shippensburg University. http://www.ship.edu/~cgboeree/maslow.html.

Bok, Derek (2003). *Universities in the Marketplace: The Commercialization of Higher Education.* Princeton, NJ: Princeton University Press.

Bollag, Burton (2005). "Congress Stops Giving Heartburn to Accreditors." *Chronicle of Higher Education* 50 (16): A1.

Bonwell, Charles C., and James A. Eison (1991). "Active Learning: Creating Excitement in the Classroom." Washington D.C.: ERIC Clearinghouse on Higher Education. www.ed.gov/databases/ERIC_Digests/ed340272.html.

Bower, Beverly L., and Kimberly P. Hardy, eds. (2005). *From Distance Education to E-Learning: Lessons along the Way.* San Francisco: Jossey-Bass.

Braskamp, Larry A., and David C. Braskamp (1997). "The Pendulum Swing of Standards and Evidence." *The CHEA Chronicle* 1 (5): 1–9. www.chea.org/Chronicle/-vol1/no5/index.cfm.

Brewer, Jerrilyn A., David Trites, Ron Matuska, and Jane Bishop (1999). "A Partnership Worth Pursuing: NCA Accreditation and the Malcolm Baldrige Award." In *A Collection of Papers on Self-Study and Institutional Improvement.* Chicago, IL: North Central Association of Colleges and Schools.

Brown, Mark G. (1997). *Baldrige Award Winning Quality: How to Interpret the Baldrige Criteria for Performance Excellence.* Milwaukee, WI: ASQ Press.

Browne, Laura D., and Cindy Green (2001). "AQIP as a Change Agent in Higher Education: Lessons Learned by a Charter Institution." In *A Collection of Papers on Self-Study and Institutional Improvement.* Chicago, IL: The Higher Learning Commission of the North Central Association of Schools and Colleges.

Burd, Stephen (2005). "Colleges Are Urged to Remain Open to Tracking Students' Progress." *Chronicle of Higher Education* 51 (31): A21.

Burke, Joseph C. (2005). "The Many Faces of Accountability." In *Achieving Accountability in Higher Education,* ed. Joseph C. Burke and Associates, San Francisco: Jossey-Bass.

Calhoun, Marry Odile (2003). "Accreditation on Planning Question" (e-mail, January 27).

Cahoon, Mary Odile, and Daniel H. Pilon (1994). "Linking the NCA Self-Study to an Annual Planning Process." In *A Collection of Papers on Self-Study and Institutional Improvement*. Chicago, IL: North Central Association of Colleges and Schools.

Calhoun, Craig (1999). "The Changing Character of College: Institutional Transformation in American Higher Education." In *The Social Worlds of Higher Education: Handbook for Teaching in a New Century*, ed. Bernice A. Pescosolido and Ronald Aminzade. Thousand Oaks, CA: Pine Forge Press.

Capen, Samuel P. (1939). *Seven Devils in Exchange for One*. Series 1. vol. 3. Washington, D.C.: American Council on Education.

Capowski, G. (1994). "Anatomy of a Leader: Where Are the Leaders of Tomorrow?," *Management Review* (March): 12–15.

Caravatta, Michael (1997). "Conducting an Organizational Self-Assessment using the 1997 Baldrige Award Criteria." *Quality Progress* 30 (10): 87–91.

Carneval, Dan (2004). "Distance Education: Keeping Up With Exploding Demand." *Chronicle of Higher Education* 50 (21): B8.

Carroll, Marilyn Nelson (2003). "Initiating the Self-Study Process: Practical Suggestions." In *A Collection of Papers on Self-Study and Institutional Improvement: The Self-Study Process for Commission Evaluation*, vol. 4, ed. Susan E. Van Kollenberg. Chicago, IL: The Higher Learning Commission of the North Central Association of Schools and Colleges.

Casey, Robert J., and John W. Harris (1979). *Accountability in Higher Education: Forces, Counterforces, and the Role of Institutional Accreditation*. Washington, D.C.: Council on Postsecondary Accreditation. www.chea.org.

Cerbo, Samuel C. (2003). *Modern Management*. 9th ed. Upper Saddle River, NJ: Prentice Hall.

Chaffee, Ellene Earle (1984). "Successful Strategic Management in Small Private Colleges," *Journal of Higher Education* 55 (2): 212–241.

Chambers, Charles M. (1983). "Council on Postsecondary Education." In *Understanding Accreditation*, ed. Arthur K. Yeung, Charles M. Chambers, and H. R. Kells. San Francisco, CA: Jossey-Bass.

Chapman, Alan (2005). *Douglas McGregor—Theory X and Theory Y*. www.businessballs.com. http://www.businessballs.com/mcgregor.htm

CHEA (1999). *CHEA Almanac of External Quality Review*. Washington, D.C.: Council for Higher Education Accreditation. www.chea.org.

——— (2001). "CHEA's Recognition Policy and Procedures." Council for Higher Education Accreditation. http://www.chea.org/About/Recognition.cfm.

——— (2002). "Directory of CHEA Participating and Recognized Organizations 2001–2002," vol. 2002. Council for Higher Education Accreditation.

——— (2004a). "CHEA Update." Council for Higher Education Accreditation. www.chea.org/Government/CHEA_HEA13.htm.

——— (2004b). "New HEA Bill Creates Strong Reaction." Council for Higher Education Accreditation. www.chea.org/Government/HEAupdate/CHEA_HEA11.htm.

Claerhout, Lori-Ann and Peter S. Cookson (2000). "An International Benchmarking Study of Copyright Operations in Distance Education Universities." *Journal of Distance Education* 15 (2), 85–96.

Clark, Burton (1963). "Faculty Culture." In *The Study of Campus Cultures*, ed. T. Lunsford. Boulder, CO: Western Interstate Commission on Higher Education.

——— (1989). "The Academic Life: Small Worlds, Different Worlds." *Educational Researcher* 18 (5): 4–8.

CQIN (2003). "Educational Excellence." *Continuous Quality Improvement Network*. http://www.cqin.org/educationalexcellence/pacesetter_award.asp.

CRAC (1999). "Proposed Policies on the Accreditation of Institutions Operating Interregionally." NCA Higher Learning Commission. http://www.ncahigherlearningcommission.org/index.php?option=com_content&task=view&id=50&Itemid=85.

Davenport, Cynthia A. (2001). "How Frequently Should Accreditation Standards Change?" In *How Accreditation Influences Assessment*, ed. James L. Ratcliff, Edward S. Lubinescu, and Maureen A. Gaffney. San Francisco: Jossey-Bass.

Davies, Peter, ed. (1981). *The American Heritage Dictionary of the English Language*. 3rd ed. New York: Dell Publishing.

Degree.net (2000a). "Accreditation." Degree.net. http://www.degree.net/guides/accreditation_guide.html.

——— (2000b). "Accrediting Agencies Not Recognized Under GAAP." Degree.net. ww.degree.net/guides/non-gaap_listings.html

DETC (2004). "DETC Accreditation Overview." Accrediting Commission of the Distance Education and Training Council. www.detc.org.

Dill, David D. (2000). "Is There an Academic Audit In Your Future?" *Change* 32. www.findarticles.com/cf_0/m1254/4_32/64189900/print.jhtml.

Dill, David D., William F. Massy, and Peter R. Williams (1996). "Accreditation and Academic Quality Assurance: Can We Get There From Here?" *Change* 28 (5): 16–24.

Dill, William R. (1998). "Specialized Accreditation: An Idea Whose Time Has Come?" *Change* (July–August): 18–25.

Doerr, Arthur H. (1983). "Accreditation — Academic Boon or Bane." *Contemporary Education* 55 (1): 6–8.

Dohman, Gloria, and Harvey Link (2001). "Understanding Your Campus Culture is Key to Self-Study Process." In *A Collection of Papers on Self-Study and Institutional Improvement*. Chicago, IL: North Central Association of Colleges and Schools Commission on Institutions of Higher Education.

Drucker, Peter F. (1993). *Post Capitalist Society*. New York: HarperCollins.

——— (2001). *The Essential Drucker: In One Volume the Best of Sixty Years of Drucker's Essential Writings on Management*. New York: HarperBusiness.

Duderstadt, James J., and Farris W. Womack (2003). *The Future of the Public University in America*. Baltimore, MD: Johns Hopkins University Press.

Eaton, Judith S. (1999a). "Advancing Quality through Additional Attention to Results." *CHEA Chronicle* 1 (11): 1–8. www.chea.org/Chronicle/vol1/no11/index.cfm.

——— (1999b). "Letter from the President." Council for Higher Education Accreditation. www.chea.org/Research/president-letters/99-07-16.cfm.

——— (2001a). "Regional Accreditation Reform." *Change* 33 (2): 39–45.

——— (2001b). "Taking a Look at Ourselves, Accreditation." In *Enhancing Usefulness*. Chicago, IL: Council for Higher Education. www.chea.org/Research/president-letters/01-08.cfm.

——— (2002). *Maintaining the Delicate Balance: Distance Learning, Higher Education Accreditation, and the Politics of Self-Regulation*. Washington, D.C.: American Council on Education.

——— (2003a). "Before You Bash Accreditation: Consider the Alternatives." *Chronicle Review*. Washington, D.C.: Chronicle of Higher Education, 49 (25): B15.

——— (2003b). "The Value of Accreditation: Four Pivotal Roles." In *CHEA Letter from the President*. Council for Higher Education Accreditation.

——— (2004). "Is the Era of Self-Regulation Over?" In *CHEA Letter from the President*. Council for Higher Education Accreditation.

Eggers, Walter (2000). "The Value of Accreditation in Planning." *CHEA Chronicle* 3 (1): 1–7. www.chea.org/Chronicle/vol3/no1/value.cfm.

El-Khawas, Elaine (2001). *Accreditation in the USA: Origins, Developments and Future Prospects*. Paris: Institute for Educational Planning—UNESCO. www.unesco.org/iiep

Elman, Sandra E. (1994). "Academic Freedom and Regional Accreditation: Guarantors of Quality in the Academy." *New Directions for Higher Education* 88 (Winter): 89–100.

Elmuti, Dean (1997). "The Perceived Impact of Team-Based Management Systems on Organizational Effectiveness." *Team Performance Management: An International Journal* 3 (3): 179–192.

Ewell, Peter T. (1994). "Accountability and the Future of Self-Regulation: A Matter of Integrity." *Change* 26 (6): 24–29.

——— (1998). "Examining a Brave New World: How Accreditation Might Be Different." In CHEA "Usefulness" Conference. Washington, D.C.: Council for Higher Education Accreditation. www.chea.org/Events/Usefulness/98May/98_05Ewell.html.

——— (2001). "Toward Excellence in Learning." Paper presented at the Annual Conference of the Middle States Commission on Higher Education. Baltimore, MD.

Farrell, Elizabeth F. (2003). "A Common Yardstick?" *Chronicle of Higher Education,* 49 (49): A25.

Fayol, Henri (1930). *Industrial and General Administration.* Trans. J. A. Coubrough. Geneva: International Management Institute.

Feddersen, Bill (1999). "Use of Baldrige Criteria for an Accreditation Self-Study." In *A Collection of Papers on Self-Study and Institutional Improvement.* Chicago, IL: North Central Association of Colleges and Schools.

Fideler, David (1996). "The Seven Liberal Arts." Cosmopolis.com. http://cosmopolis.com/villa/liberal-arts.html.

Foster, Andrea L. (2002). "Plagiarism-Detection Tool Creates Legal Quandary." *Chronicle of Higher Education* 48 (36): A37.

George, S. (1992). *The Baldrige Quality System: The Do-It Yourself Way to Transform your Business.* New York: Wiley.

Geroge, C. S., Jr. (1972). *The History of Management Thought.* Englewood Cliffs, NJ: Prentice Hall.

Giacomelli, Marie A. (2002). "Self-Study: The Proof is in the Plan, Process, and Product." In *A Collection of Papers on the Self-Study and Institutional Improvement.* Chicago, IL: The Higher Learning Commission of the North Central Association of Colleges and Schools.

Glidden, Robert (1996). "Accreditation at the Crossroads." *Educational Record* 77 (4): 22–24.

——— (2003). "Accreditation at a Crossroads." Council for Higher Education Accreditation. www.chea.org/Research/crossroads.cfm.

——— (2004). "Positioning Accreditation for the Future: Change or Status Quo." Paper presented at the CHEA Annual Conference. Marina del Ray, CA.

Goncalves, Karen (1992). "Those Persons Who Do Your Planning." *Planning for Higher Education* 20 (2): 25–29.

Goodman, Glay, and Rene Willekens (2001). "Self-Study and Strategic Planning: A Critical Combination." In *A Collection of Papers on Self-Study and Institutional Improvement.* Chicago, IL: North Central Association of Colleges and Schools Commission on Institutions of Higher Education.

Goodstein, Leonard D. (1978). "Organization Development in Bureaucracies: Some Caveats and Cautions." In *The Cutting Edge: Current Theory and Practice in Organization Development,* ed. W. Warner Burke. La Jolla, CA: University Associates.

Gose, Ben (2002). "A Radical Approach to Accreditation: Can the Academic Quality Improvement Project Make the Process Worth the Time and Money?" *Chronicle of Higher Education* 49 (10): A25.

Government, U.S. (2002). "Accreditation in the U.S.": Office of Postsecondary Education. http://www.ed.gov/offices/OPE/accreditation/accredus.html.

Graham, Patricia A., Richard W. Lyman, and Martin Trow (1995). *Accountability of Colleges and Universities: An Essay.* New York: Columbia University Press.

Gulick, Luther Halsey, and L. Urwick (1937). *Papers on the Science of Administration*. New York: Institute of Public Administration, Columbia University.

Hamel, Gary, and C. K Prahalad (1994). *Competing for the Future*. Boston, MA: Harvard Business School Press.

Hamm, Michael S. (1997). *The Fundamentals of Accreditation*. Washington, D.C.: American Society of Association Executives.

Harcleroad, Fred F. (1980). *Accreditation: History, Process, and Problems*. Washington, D.C.: ERIC Clearinghouse on Higher Education.

Hardin, Lynn, Susan Lucas, Linda North, and Lucille Prewitt (2005). "From Institution to Individual: The Beginnings of Organization Theory." www.susanlucas.com. susanlucas.com/it/ael682/paper.html.

Harrington, Charles F., Robert L. Reid, and Kevin J. G. Snider (2000). "Developing Data Resources for Facilitating Institutional Self-Study." In *A Collection of Papers on Self-Study and Institutional Improvement*. Chicago, IL: Commission on Institutions of Higher Education of the North Central Association of Colleges and Schools.

Henninger, Edward A. (1998). "Dean's Role in Change: The Case of Professional Accreditation in Reform of American Education Collegiate Business Education." *Journal of Higher Education Policy and Management* 20 (2): 203–213.

Henry, J., and M. Hartzler (1998). *Tools for Virtual Teams—A Team Fitness Comparison*. Milwaukee, WI: ASQ Quality Press.

Hersey, Paul, and Kenneth Blanchard (1988). *Management and Organizational Behavior*. Englewood Cliffs, NJ: Prentice Hall.

Hoffman, G. (1994). *The Technology Payoff*. Burr Ridge, IL: Irwin.

Hogg, Edward E. (1993). "Unraveling the Accreditation Enigma: A Historical Approach." (PhD thesis, The Fielding Institute).

Hutchinson, Peter M. (1994). "Building a Self-Study around Special Emphasis." In *A Collection of Papers on Self-Study and Institutional Improvement*. Chicago, IL: North Central Association of Colleges and Schools.

International Handbook of Universities Yearbook (2005). New York: Palgrave.

Janosik, S., Creamer D., and M. Alexander (2001). *International Perspectives on Quality in Higher Education*. Blacksburg, VA: Virginia Tech.

Jones, Deb, and Nancy Schendel (2000). "Fostering Employee Ownership in the Self-Study Process Through Committee Empowerment." In *105th Annual Meeting of the North Central Association Commission on Institutions of Higher Education*, vol.4, ed. Susan E. Van Kollenburg. Chicago, IL: North Central Association Commission on Institutions of Higher Education.

Jones, Gareth R., and Jennifer M. George (2003). *Contemporary Management*. New York: McGraw-Hill.

Jones, L. R., Fred Thompson, and William Zumeta (2001). "Public Management for the New Millennium: Developing a Relevant and Integrated Professional Curricula?" International Public Management Network. http://www.inpuma.net/test2/issue3/PMCURR8.1doc.pdf.

Kells, H. R. (1994). *Self-Study Processes: A Guide for Postsecondary and Similar Service-Oriented Institutions and Programs*. 3rd ed. Greenwood, CT: Oryx Press.

Kells, H. R., and Robert Kirkwood (1979). "Institutional Self-Evaluation Processes." *Educational Record* (Winter): 25–45.

Kemling, Katherine (1994). "The Two Elements That Made All the Difference." In *A Collection of Papers on Self-Study and Institutional Improvement*. Chicago, IL: North Central Association of Colleges and Schools.

Kezar, Adrianna J. (2000). "Higher Education Trends." Vol. 2002. Washington, D.C.: ERIC Clearinghouse on Higher Education. www.eriche.org/trends/teaching.html.

——— (2004). "Obtaining Integrity? Reviewing and Examining the Charter Between Higher Education and Society." *Review of Higher Education* 27 (4): 429–459.

Kotter, J. P. (1990a). *A Force for Change: How Leadership Differs from Management.* New York: Free Press.

——— (1990b). "What Leaders Really Do." *Harvard Business Review* (May–June): 103–111.

Lawrence, John J., and Byron Dangerfield (2001). "Integrating Professional Reaccreditation and Quality Award Processes." *Quality Assurance in Education* 9 (2): 80–91.

Leef, George C., and Roxana D. Burris (2002). *Can College Accreditation Live Up to its Promise?* Washington, D.C.: American Council of Trustees and Alumni. www.goacta.org.

LeMaster, Brenda, and Lawrence Johnson (2001). "Electronic Data Gathering in the Preparation of the Self-Study." In *A Collection of Papers on Self-Study and Institutional Improvement*. Chicago, IL: North Central Association of Colleges and Schools Commission on Institutions of Higher Education.

Lenn, Marjorie Peace, and Lenora Campos (1997). *Globalization of the Professions and the Quality Imperative: Professional Accreditation, Certification and Licensure*. Madison, WI: Magna Publications.

Lindquist, Sarah B. (2004). "Using a Virtual Data and Documents Center to Support the Self-Study, Visit and Beyond." In *109th Annual Meeting of the North Central Association of Schools and Colleges*, ed. Susan E. Van Kollenberg. Chicago, IL: The Higher Learning Commission.

——— (2005). "Using a Virtual Resource Center to Support the Self-Study, Visit, and Beyond." In *110th Annual Meeting of the North Central Association*, ed. Susan E. Van Kollenberg. Chicago, IL: The Higher Learning Commission of the North Central Association of Colleges and Schools.

Lingenfelter, Paul E., and Charles S. Lenth (2005). "What Should Reauthorization Be About?" *Change* 37 (3): 12–19.

Lynn, Laurence E., Jr., and Sydney Stein, Jr. (2003). "Public Management." In *Handbook of Public Administration*, ed. B. Guy Peters and Jon Pierre. London: Sage Publications.

Maher, John (2000). *A Total Quality Strategic Planning Database Model Linking Objectives with Performance Measures and Accreditation Criteria*. Virginia Beach, VA: Regent University.

Martin, Jeffrey C. (1994). "Recent Developments Concerning Accrediting Agencies in Postsecondary Education." *Law and Contemporary Problems* 57 (4): 121–149.

Martin, Rebecca R., Kathleen Manning, and Judith A. Ramaley (2001). "The Self-Study as a Chariot for Strategic Change." *New Directions for Higher Education* (113): 95–115.

Massy, William F., and Robert Zemsky (2004). "Thwarted Innovation: What Happened to E-Learning and Why," The Learning Alliance for Higher Education. http://www.irhe.upenn.edu/WeatherStation.html.

——— (2004). "Thwarted Innovation? The Research Says What It Says." *Quarterly Review of Distance Education* 5 (4): xi–xii.

Masters, Brenda (2004). "Accreditation Web Site: A Process Vehicle for the Self-Study." In *109th Annual Meeting of the North Central Association of Schools and Colleges*, vol. 4, ed. Susan E. Van Kollenberg. Chicago, IL: The Higher Learning Commission.

McCallin, Rose C. (1999). "The Accreditation Process: Connecting Potential Disconnects on Campus." In *A Collection of Papers on Self-Study and Institutional Improvement*. Chicago, IL: North Central Association of Colleges and Schools Commission on Higher Education.

McGregor, Douglas (1960). *The Human Side of Enterprise*. New York: McGraw-Hill.

McGuinness, Aims C. (2002). "Reflections on Postsecondary Governance Changes." Education Commission of the States. http://www.ecs.org/clearinghouse/37/76/3776.htm.

Meyer, Herbert H., Emanuel Kay, and John R. P. French, Jr. (1965). "Split Roles in Performance Appraisal." *Harvard Business Review* 43: 123–129.

Michels, Jennifer (1998). "ARC's Corporate Accreditation Plan is Slow Out of the Blocks." In *Travel Agent*. September 5.

Michigan (2005). "Unapproved Accrediting Bodies." State of Michigan. http://www.michigan.gov/documents/Non-accreditedSchools_78090_7.pdf .

Miller, Danny (1993). "The Architecture of Simplicity." *Academy of Management Review* (January): 116–138.

Mintzberg, Henry (1994). *The Rise and Fall of Strategic Planning*. New York: Free Press.

Moore, Michael R., and Michael A. Diamond (2000). *Academic Leadership: Turning Vision into Reality*. Cleveland, OH: The Ernst & Young Foundation.

Morgan, Richard (2002). "Lawmakers at Hearing on College-Accreditation System Call for More Accountability." *Chronicle of Higher Education*. chronicle.com/daily/2002/10/2002100204n.htm.

MSACHE (1995). "Use and Distribution of Evaluation Team Reports." Middle States Commission on Higher Education. www.msache.org/Designs.pdf.

——— (2001). "Follow-up Reports and Visits." Middle States Commission on Higher Education. www.msache.org/pofolre.html.

——— (2002). "Designs for Excellence." Middle States Commission on Higher Education. www.msache.org/Designs.pdf.

NASC (2002a). "Evaluation Committee." Commission on Colleges and Universities of the Northwest Association of Schools and Colleges and Universities. http://www.nwccu.org/.

——— (2002b). "Institutional Commitment and Responsibilities." Commission on Colleges of the Northwest Association of Schools and Colleges. http://www.nwccu.org/.

Natale, Samuel M., Anthony F. Libertella, and Barbara Edwards (1998). "Team Management: Developing Concerns." *Team Performance Management: An International Journal* 4 (8): 319–330.

NCA (2001a). "Academic Quality Improvement Project." in *Synthesis: An Annual Publication of the Higher Learning Commission*. Chicago, IL: The Higher Learning Commission of the North Central Association of Colleges and Schools.

——— (2001b). "Statement of Commitment by the Regional Accrediting Commissions for the Evaluation of Electronically Offered Degree and Certificate Programs." Council of Regional Accrediting Commissions. www.ncahlc.org.

——— (2003). "Institutional Accreditation: An Overview." The Higher Learning Commission of the North Central Association of Colleges and Schools. www.ncahlc.org.

——— (2006). "Guidelines for Distance Education." : The Higher Learning Commission of the North Central Association of Colleges and Schools. www.ncahlc.org.

Neelankavil, J. P. (1994). "Corporate America's Quest for an Ideal MBA." *Journal of Management Development* 13 (5): 38–52.

Newton, Robert (1992). "The Two Cultures of Academe: An Overlooked Planning Hurdle." *Planning for Higher Education* 21 (1): 8–14.

NIST (1987). "Baldrige National Quality Program." National Institute of Standards and Technology. http://www.quality.nist.gov/Improvement_Act.htm.

——— (2001a). "2001 Malcolm Baldrige National Quality Award Education Category." . National Institute for Standards and Technology. www.nist.gov/public_affairs/factsheet/education.htm.

——— (2001b). "President and Commerce Secretary Announce Recipients of Nation's Highest Honor in Quality and Performance Excellence." National Institute of Standards and Technology. http://www.nist.gov/public_affairs/releases/g01-110.htm.

Noonan, Norma C., and Kathryn Heltne Swanson (2003). "Lessons Learned from Years of Reviewing Self-Study Reports: Advice on Writing and Editing the Study." In *A Collection of Papers on Self-Study and Institutional Improvement: The Self-Study Process for Commission Evaluation,* vol. 4, ed. Susan E. Van Kollenberg. Chicago, IL: The Higher Learning Commission of the North Central Association of Schools and Colleges.

NWCCU (2005). "Decisions." Northwest Commission on Colleges and Universities. http://www.nwccu.org/Process/Decisions/Decisions.htm.

Octogram (2003). "The Dean's Associate." Octogram, Inc. www.octogram.com.

Orlans, Harold (1980). "The End of a Monopoly." *Change* 12.

Palloff, Rena M., and Keith Pratt (2004). *Collaborating Online: Learning Together in Community.* San Francisco: Jossey-Bass.

Palmer, Emma (2002). "Strategies for a Productive Team Visit." In *A Collection of Papers on Self-Study and Institutional Improvement.* Chicago, IL: The Higher Learning Commission of the North Central Association of Colleges and Schools.

Peters, Thomas J. (1992). *Liberation Management: Necessary Disorganization for the Nanosecond Nineties.* New York: Alfred A. Knopf.

Pfnister, A. O. (1959). "Accreditation in the North Central Region." In *Accreditation in Higher Education.* Washington, D.C.: Office of Education, U.S. Department of Health, Education, and Welfare.

——— (1971). "Regional Accrediting at a Crossroads." *Journal of Higher Education* 47 (7): 558–573.

Porter, Lyman W., and Lawrence E. McKibbin (1988). *Management Education and Development: Drift or Thrust into the 21st Century.* New York: McGraw-Hill.

PRNEWSWIRE (2000). "GeoAccess and CSI Advantage Partner to Offer Accreditation Management Software AIMS Software Reduces NCQA Survey Preparation Time By 50 Percent." PR Newswire Association, Inc. www.prnewswire.com.

Pugh, D. (1990). *Organization Theory.* Baltimore: Penguin Press.

Raynor, William III (1999). "Business Accreditation: To Enhance Global Labor Standards." *Humanist* 59. www.findarticles.com/cf_0/m1374/2_59/54099127/print.jhtml.

Reed, Ronald D., and Rolf C. Enger (2000). "Team Building: Key to a Successful Self-Study and Team Visit." In *A Collection of Papers on Self-Study and Institutional Improvement.* Chicago, IL: Commission on Institutions of Higher Education of the North Central Association of Schools and Colleges.

Register, Federal (1995). "Request for Comments on Agencies Applying to the Secretary for Initial Recognition or Renewal of Recognition." U.S. Department of Education. http://www.ed.gov/legislation/FedRegister/announcements/1995-4/fr21de5c.html.

Reich, Michael J. (1999). "Question: What Worked Best?" In *A Collection of Papers on Self-Study and Institutional Improvement.* Chicago, IL: Committee on Institutions of Higher Education of the North Central Association of Schools and Colleges.

Rieves, Nancy (1999). "A Self-Study Coordinator's Responsibility: From Campus-Wide Involvement to Community Support." In *104th Annual Meeting of the North Central Association Commission on Institutions of Higher Education,* vol. 4, ed. Susan E. Van Kollenburg. Chicago, IL: North Central Association Commission on Institutions of Higher Education.

Robbins, Stephen P. (2000). *Managing Today.* 2nd ed. Upper Saddle River, NJ: Prentice Hall.

Robinson-Weening, Lynette (1995). "A Study of the Relationship between Regional Accreditation and Institutional Improvement among New England Colleges and Universities" (PhD thesis, at Boston College).

Rudolph, Frederick (1977). *Curriculum: A History of the American Undergraduate Course of Study Since 1636.* San Francisco: Jossey-Bass.

Russell, J. D., and C. H. Judd (1940). *The American Educational System.* Cambridge: Riverside Press.

SACS (2002). "General Information on the Accreditation Process." Commission on Colleges of the Southern Association of Colleges and Schools. http://www.sacscoc.org/genac-cproc.asp.

Safman, Phyllis C. (1998). "What Accreditors Say: Changes in Approach to Accreditation Practice." *CHEA Chronicle* 1 (10): 1–3. www.chea.org/Chronicle.vol1/no10/practices.cfm.

Sanders, Diana W. and Michael D. Richardson (2002). "Whose Property Is It Anyhow? Using Electronic Media in the Academic World." *Journal of Technology Studies* 28 (2): 117–123.

Schermerhorn, John R. (2002). *Management.* 7th ed. New York: Wiley.

Scott, Kenneth, and Allan Walker (1999). "Extending Teamwork in Schools: Support and Organisational Consistency." *Team Performance Management: An International Journal* 5 (2): 50–59.

Selden, William K. (1960). *Accreditation: A Struggle over Standards in Higher Education.* New York: Harper & Brothers.

Selingo, Jeffrey (2004). "U.S. Public's Confidence in Colleges Remains High." *Chronicle of Higher Education* 50 (35): A1.

Semrow, Joseph J., Joseph A. Barney, Marcel Fredericks, Janet Fredericks, Patricia Robinson, and Allan O. Pfnister (1992). *In Search of Quality: The Development, Status and Forecast of Standards in Postsecondary Accreditation.* New York: Peter Lang.

Seymour, Daniel (1996). *High Performing Colleges: Volume Two: Case and Practice.* Maryville, MO: Prescott Publishing Company.

Seymour, Daniel, and Satinder K. Dhiman (1996). "Baldrige Barriers." In *High Performing Colleges,* vol.1, *Theory and Concepts,* ed. Daniel Seymour. Maryville, MO: Prescott Publishing Company.

Shafritz, Jay M., and Jay Steven Ott (2001). *Classics of Organization Theory.* Philadelphia: Harcourt Brace.

Shaw, Robert (1993). "A Backward Glance: To a Time before There Was Accreditation." *NCA Quarterly* 68 (2): 323–335.

Simon, Herbert A., Victor Thompson, and Donald W. Smithburg (1991). *Public Administration.* New Brunswick: Transaction Publishers.

Simons, Robert (1995). *Levers of Control: How Managers Use Innovative Control Systems to Drive Strategic Renewal.* Boston: Harvard Business School Press.

Simonson, Michael (2004). "'Thwarted Innovation' or Thwarted Research?" *Quarterly Review of Distance Education* 5 (4): vii–ix.

Sorenson, Charles W., Julie A. Furst-Bowe, and Carol T. Mooney (2002). "Lessons for Higher Education Planning: Applying the Baldrige Criteria." In *A Collection of Papers on Self-Study and Institutional Improvement.* Chicago, IL: The Higher Learning Commission of the North Central Association of Colleges and Schools.

Stewart, George R., and Brian H. Kleiner (1995). "The Enabling Power to Teams and Information Technology." *Team Performance Management: An International Journal* 2 (2): 13–18.

Swanson, Kathryn Heltne (1999). "Data ARE Practicalities Relevant to Writing the Self-Study." In *A Collection of Papers on Self-Study and Institutional Improvement.* Chicago, IL: North Central Association of Colleges and Schools Commission on Higher Education.

Swanson, Ronald A., Helene Hedlund, and Jerry L. Neff (2000). "Incorporating the Malcolm Baldrige National Quality Criteria into Your NCA Accreditation Process." In *A Collection of Papers on Self-Study and Institutional Improvement.* Chicago, IL: North Central Association of Colleges and Schools.

Taylor, Mark (2002). "Managing the Stress of the Self-Study Process." In *A Collection of Papers on Self-Study and Institutional Improvement*. Chicago, IL: The Higher Learning Commission of the North Central Association of Colleges and Schools.

Thompson, Hugh L. (1993). "Accreditation: Recharting the Future of Accreditation." *Educational Record* 74 (4): 39–42.

Tobin, Ronald W. (1994). "The Age of Accreditation: A Regional Perspective." *Academe* 80 (4): 26–33.

Tompkins, E., and W. H. Gaumnitz (1954). *The Carnegie Unit: Its Origins, Status and Trends*. Washington, D.C.: U.S. Government Printing Office.

Trow, Martin (1973). *Problems in the Transition from Elite to Mass to University Higher Education*. Berkeley, CA: Carnegie Commission on Higher Education.

——— (1994). "Managerialism and the Academic Profession: The Case of England." *Higher Education Policy* 7 (2): 11–18.

——— (1996). "Trust, Markets, and Accountability in Higher Education: A Comparative Perspective." *Higher Education Policy* 9 (4): 309–324.

UMICH (2000). "A Report of a Focused Visit to University of Michigan–Ann Arbor." Commission on Institutions of Higher Education of the North Central Association of Colleges and Schools. http://www.umich.edu/~provost/slfstudy/pdf/accredreport.pdf.

UNESCO (2004). "List of Universities of the World." International Association of Universities. http://www.unesco.org/iau/onlinedatabases/list.html.

USDE (2006). "College Accreditation in the United States." U.S. Department of Education. http://www.ed.gov/admins/finaid/accred/accreditation_pg6.html#NationallyRecognized.

Vaughn, John (2002). "Accreditation, Commercial Rankings, and New Approaches to Assessing the Quality of University Research and Education Programmes in the United States." *Higher Education in Europe* 27 (4): 433–441.

Vergari, Sandra, and Frederick M. Hess (2004). "The Accreditation Game." *Education Next* 2 (3): 48–57.

Veysey, Laurence R. (1965). *The Emergence of the American University*. Chicago, IL: University of Chicago Press.

Walck, Christa (1998). "Organizing and Selling the Self-Study Process to the University Community." In *103rd Annual Meeting of the North Central Association of Colleges and Schools*, vol. 4, ed. Susan E. Van Kollenberg. Chicago, IL: Commission on Institutions of Higher Education, North Central Association of Colleges and Schools.

Watts, Margit Misngyi, ed. (2003). *Technology: Taking the Distance out of Learning: New Directions for Teaching and Learning*. San Francisco: Jossey-Bass.

WCET (2006). "Western Cooperative for Educational Telecommunications." Vol. 2006: Western Interstate Commission for Higher Education. http://www.wcet.info/about/.

Webster (2002). *Webster's Third New International Directory*. Springfield, MA: Merriam-Webster.

Wikipedia (2006). "List of Unrecognized Accreditation Associations of Higher Learning." http://en.wikipedia.org/wiki/List_of_unrecognized_accreditation_associations_of_higher_learning.

Wilkins, Theresa B. (1959). "Accreditation in the States." In *Accreditation in Higher Education*, ed. Lloyd E. Blanch. Washington, D.C.: U.S. Department of Health, Education, and Welfare, Office of Education.

Winona (2003). "The Special Emphasis Study Option from the NCA Handbook of Accreditation." Winona State University. http://www.winona.edu/air/nca/New_Folder/ncasproposal.htm.

www.aju.edu (2006). "Accrediting Bodies NOT Recognized by the United States Department of Education." Andrew Jackson University. http://www.aju.edu/usdoe_faqs.htm.

Yee, Carole, and Scott Zeman (2005). "How We Used the New Criteria and Lived to Tell the Tale." In *110th Annual Meeting of the North Central Association of Colleges and Schools,* vol. 4, ed. Susan E. Van Kollenberg. Chicago, IL: The Higher Learning Commission.

Yeung, Arthur K., David O. Ulrich, Stephen W. Nason, and Mary Ann Von Glinow (1999). *Organizational Learning Capability: Generating and Generalizing Ideas with Impact.* Oxford: Oxford University Press.

Young, Kenneth E., Charles M. Chambers, and H. R. Kells (1983). *Understanding Accreditation.* San Francisco: Jossey-Bass.

Index

AACRAO, 21
AACSB, xiii, 30, 41, 51, 55, 58, 62, 77, 117, 139, 167
Abraham Maslow, 86
academic audit, 6, 26
academic freedom, 3, 5, 140, 176
accountability, xii, 1–2, 14, 16, 19, 23, 25–26, 32–33, 111, 119, 133, 139, 142, 147, 162, 164, 167, 172–174, 176–177
Accreditation Information Management System, 117
Accreditation Management, ix, 39, 42–43, 65, 81, 91, 129, 148
Accrediting Commission of Career Schools and Colleges of Technology, 36
Accrediting Council for Independent Colleges and Schools, 36
active-learning, 5
Adam Smith, 83, 192
Airlines Reporting Corporation, 27
American Council of Trustees and Alumni, 32
American Council on Education, 39
American Medical Association, 12, 15
AQIP, 25, 102, 133, 141–142
Arizona State University, 155
Assessment, 112, 138, 203
assessment, 2, 5, 6, 14, 23, 26, 43, 50, 53, 61, 68, 76–77, 94–95, 97–98, 100, 102–105, 111–112, 117–118, 121, 135, 137, 139, 141, 150–151, 162–163, 169–170, 200, 211
Association of American Universities, 14, 18
Australian Universities Quality Agency, 134

Baldrige, ix, xii, 24–25, 102, 133–139, 141–142
Benchmarking, xiii, 58, 136, 169
Best Practices, xiii, 168
Blackboard, 145–146, 148
Board of Regents of New York State, 13
Bowling Green State University, 105
Budgeting, 129
budgeting, xi–xii, 5, 40, 47, 85, 87, 101, 119, 129, 170
business process re-engineering, 87

Carnegie, 15, 18, 105
centralization, 2
CHEA, 7, 16–17, 19–20, 35, 37, 50, 90, 94, 160, 169, 179
Chester Bernard, 86
College Entrance Examination Board, 13
College of Financial Planning, 114
College of St. Scholastica, 96–98
Columbia University, 12, 32, 166
committees, xi, xiii, 58–59, 66, 89, 94, 97, 105, 107, 110, 115, 122–125, 130, 135, 143, 145, 147–157, 171
Congress, 2, 43
Consortium for Benchmarking for Higher Education Benchmarking Analysis, 58
continuous improvement, 119, 136–137, 141, 177
Continuous Quality Improvement Network, 26, 142
COPA, 14, 16, 19, 23
Cornell University, 166

246 • Index

CORPA, 16, 19
Council of Regional Accrediting Commissions, 167
Criteria for Accreditation, 119, 156–157, 199
crossroads, 2–4, 174

decision-making, 19, 53, 71, 98, 100–101, 111–112, 117, 168, 207–209, 211, 214
Department of Education, 16, 20, 33, 37, 167, 179, 194
Diploma mills, 37
Distance Education/Distance Learning, xii–xiii, 3, 17, 41, 148, 159–170, 190, 192, 203
diversity, 176–177

Eastern Michigan University, 109
Educational Benchmarking Incorporated, 58
e-learning systems, 60, 148–156, 168
Erie Railroad, 84
Estrella Mountain Community College, 99–100
European Association of Universities, 26

facilitation, 115, 118
Federal government, 3, 39, 174
Federation of Regional Accrediting Commissions of Higher Education, 19
financial aid, 32, 39, 170
Frederick Taylor, 84

Generally Accepted Accrediting Practices, 20, 37, 187
GeoAccess, 117
Germany, 18, 195
government, 133, 176

Harvard University, 89
Henri Fayol, 85
Henry Metcalfe, 84
Henry Poor, 84
Herbert Simon, 86
Higher Education Act, 32–33, 35–36, 43
history, xi, xii, 4, 13, 36, 83, 90, 116, 138, 162, 171

Idaho Quality Award, 139–140
Initial Accreditation, 77

Interim Report, 77
Internet, 41, 60, 78, 117, 126, 148, 153, 155–156, 169

King's College, 12

Luther Gulick, 85, 88

Management and Organization Concepts, 83
management by objectives, 87
management, 111
Maryland, 12, 166, 192–193
Michigan Tech, 57
Middle States Association, xiii, 17, 193
Minnesota, 96, 137
mission, 2–4, 14, 16, 18–19, 27, 33, 43, 47–48, 51, 54, 58, 87, 93–94, 97, 99–102, 104, 117–119, 124, 126, 142–143, 147, 163, 167, 171, 173, 175, 199–201, 209–210, 213
Moses, 83
Mt. San Antonio College, 135

National Center for Education Statistics, 162
National Commission on Accreditation, 18
National Committee for Quality Assurance, 117
National Committee of Regional Accrediting Agencies, 18
NCA, ix, 110, 114, 134, 138, 141–142, 203–204
NCATE, 117, 129
New York, xiii, 12–13, 84, 166
No Child Left Behind Act, 33
NOOSR, 21
North Central Association, 15, 17, 25, 51, 99–100, 107, 114, 138, 145, 199, 203
North Dakota State College of Science, 112
Northwest Association, 17, 204

Ohio, 105
On-Site Evaluation, 67
organizing, xi, 5, 40, 82, 85, 108–109, 113
outcomes, 118, 134, 172, 176

Periodic Review Report, 61
Pike Peak Community College, 60
planning, xi–xii, 5, 21, 24–26, 31, 40, 42–43, 47–49, 51, 57, 60–63, 66, 75, 77,

82–85, 87–90, 93–105, 107, 111–113, 115–119, 122, 129–130, 134–138, 141, 145, 150, 160, 170–172, 175–176, 200
POSDCORB, 85, 88–89
Probation, 77, 219
public opinion, 3

Reaffirm Accreditation, 77
Regent University, 118–119
regional, xii, xiii, xv, 1, 3–6, 13–18, 20–21, 25–26, 30–33, 35–37, 40–42, 47, 50–51, 53–55, 57–58, 60, 62, 67, 69, 72, 76–77, 81–82, 94–101, 105–107, 109, 111–115, 119, 122–123, 129–130, 133–139, 141, 145, 147, 155–157, 159–168, 174–175, 179, 195, 199
report, 110, 113–114, 121–122, 126, 142, 171
reporting, xi–xii, 5, 23, 26, 35, 50, 53, 73, 85, 100, 108, 114, 116, 122, 134, 139
Robert K Merton, 86
Robert Morris College, 99

self-evaluation, xv, 18, 21, 25, 31–32, 39, 41, 48–49, 51, 57–58, 60, 62, 66, 75, 94, 96–100, 103–106, 109–113, 116, 118, 121–123, 126, 130, 133–135, 137, 139–140, 142–143, 146, 155–156, 166
self-regulation, 3, 5–6, 16, 34, 172–174
self-study, 107, 109–110, 112–116, 118, 121–123, 126, 129, 135–137, 139, 171, 177
Self-Study, ix, 54, 57, 59, 65, 102, 110, 114, 121, 136, 203
Southern Association of Colleges and Schools, 17
specialized, xii–xiii, 1, 3–6, 12, 14–17, 20–21, 26, 29–31, 33, 36–37, 39–42, 47, 49–50, 55, 58, 62, 67, 76–77, 81–82, 94–95, 100–101, 105, 110–111, 115, 117, 129, 133, 139–140, 145, 149, 159–160, 164, 167, 169, 173, 175, 179
steering committee, 109–110, 113, 122
student learning outcomes, 2–3, 33, 35, 41, 47, 102, 163–164, 169, 200
SWOT, ix, 58–59, 61, 100–101

technology, xii, 1–3, 34, 58, 71, 104, 116–117, 123, 125, 130, 145–148, 151–153, 155–156, 159–160, 162, 164, 166, 168–170
The Dean's Associate, 117
The Learning House in Kentucky, 164
The Wealth of Nations, 83
Theory X and Theory Y, 86–87
third generation, 133
timeline, 113
total quality management, 87, 139
transfer credit, 3
transfer credits, 35–36
Trivium and Quadrivium, 11
Trow, 133, 143, 175
turning point, ix, 2, 4, 33–34, 36, 173

UCLA, 35
UNESCO, 20, 37, 187
United States Air Force Academy, 113
universal higher education, 36
University of Buffalo, 29
University of Cincinnati, 118
University of Michigan – Ann Arbor, 107
University of Michigan, 107, 139
University of Phoenix, 160
USDE, 16, 37, 167, 179, 205
Utah, 12, 193
UW-Stout, 134

Virginia, 12, 118
virtual resource center, 125, 151, 155, 157
virtual teams, xiii, 151
visitation team, 50, 61, 67, 70, 76, 89, 105, 130, 156
voluntary, 4, 12, 17, 27, 31–32, 35, 40, 174, 176, 206–207

Washington, 12, 20, 188–189, 193
WebCT, 148
Western Association of Schools and Colleges, 17, 195
Western Cooperative for Educational Telecommunications, 164, 167
Western Interstate Commission for Higher Education, 164, 167–168

zero-base budgeting, 87

GPSR Compliance

The European Union's (EU) General Product Safety Regulation (GPSR) is a set of rules that requires consumer products to be safe and our obligations to ensure this.

If you have any concerns about our products, you can contact us on

ProductSafety@springernature.com

In case Publisher is established outside the EU, the EU authorized representative is:

Springer Nature Customer Service Center GmbH
Europaplatz 3
69115 Heidelberg, Germany

www.ingramcontent.com/pod-product-compliance
Lightning Source LLC
LaVergne TN
LVHW041625060526
838200LV00040B/1443